COLORS FOR SURVIVAL

MIMICRY AND CAMOUFLAGE IN NATURE

COLORS FOR SURVIVAL
MIMICRY AND CAMOUFLAGE IN NATURE

Text
Marco Ferrari

Editorial Director
Valeria Manferto

Design
Patrizia Balocco

PAGE 1
*A butterfly closely resembling
a leaf rests on a blade of
grass in the South American
rainforest.*

PAGES 2-3
*A Thyria jacobeae imitates
the leaf of the arrowroot plant.*

PAGES 4-5
*A frogfish is effectively cam-
ouflaged against the corals of
the Red Sea.*

PAGES 6-7
*In Venezuela, a crocodile
blends in perfectly with the
algae on the surface of the
water.*

PAGE 8
*The carpet of leaves in the
tropical forests of Malaysia
provides the perfect habitat
for mimetic animals such as
this horned toad.*

PAGE 9
*Only the eyes of this rattle-
snake are distinguishable in
the sands of the Namibian
desert.*

PAGES 10-11
*A small cirriped mimics the
brilliant background of corals
in the Red Sea.*

PAGES 12-13
*Having a transparent body is
one way to "disappear," as
these shrimps do so effectively
against a sea anemone at
Brothers Island.*

Contents

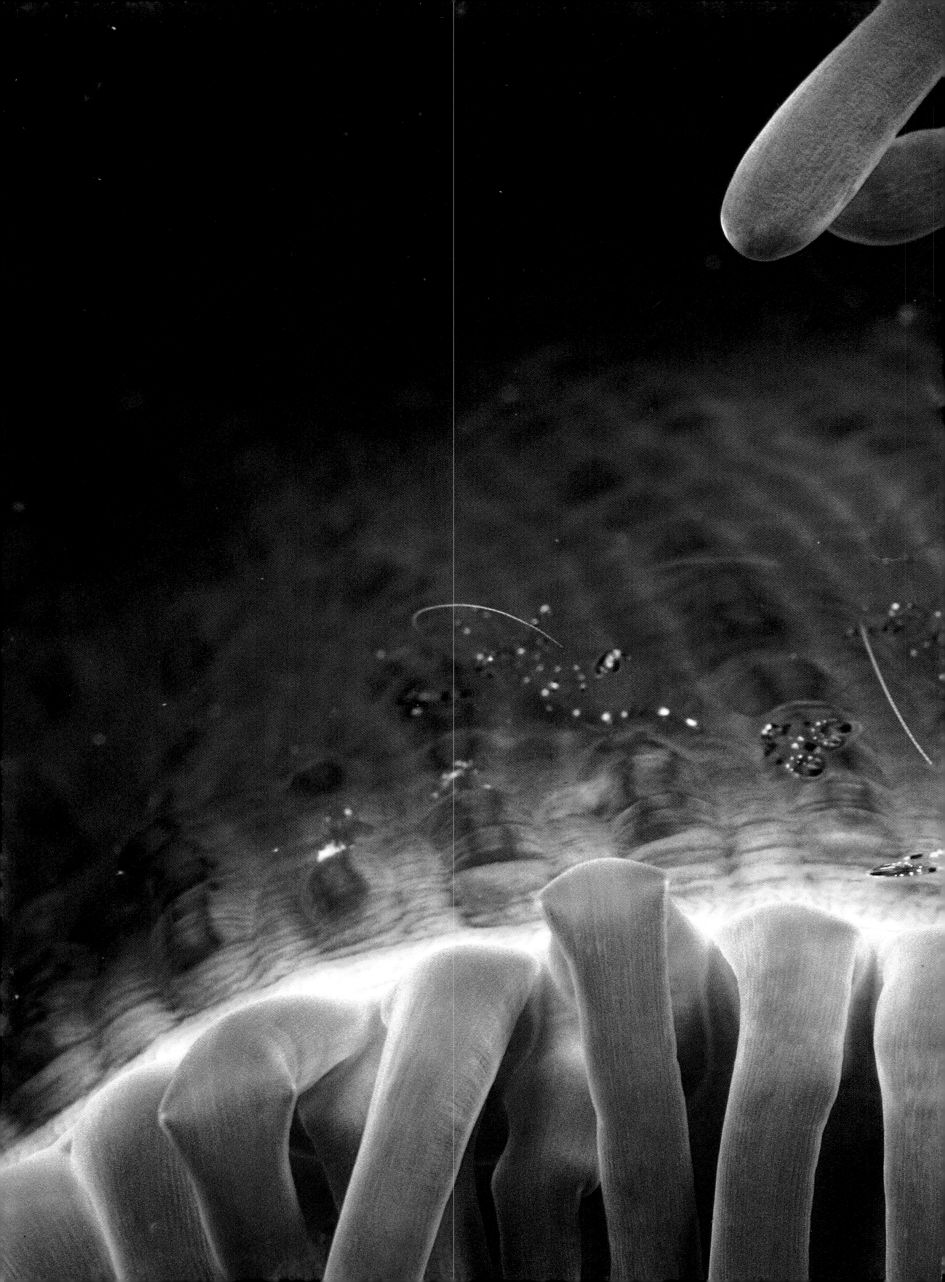

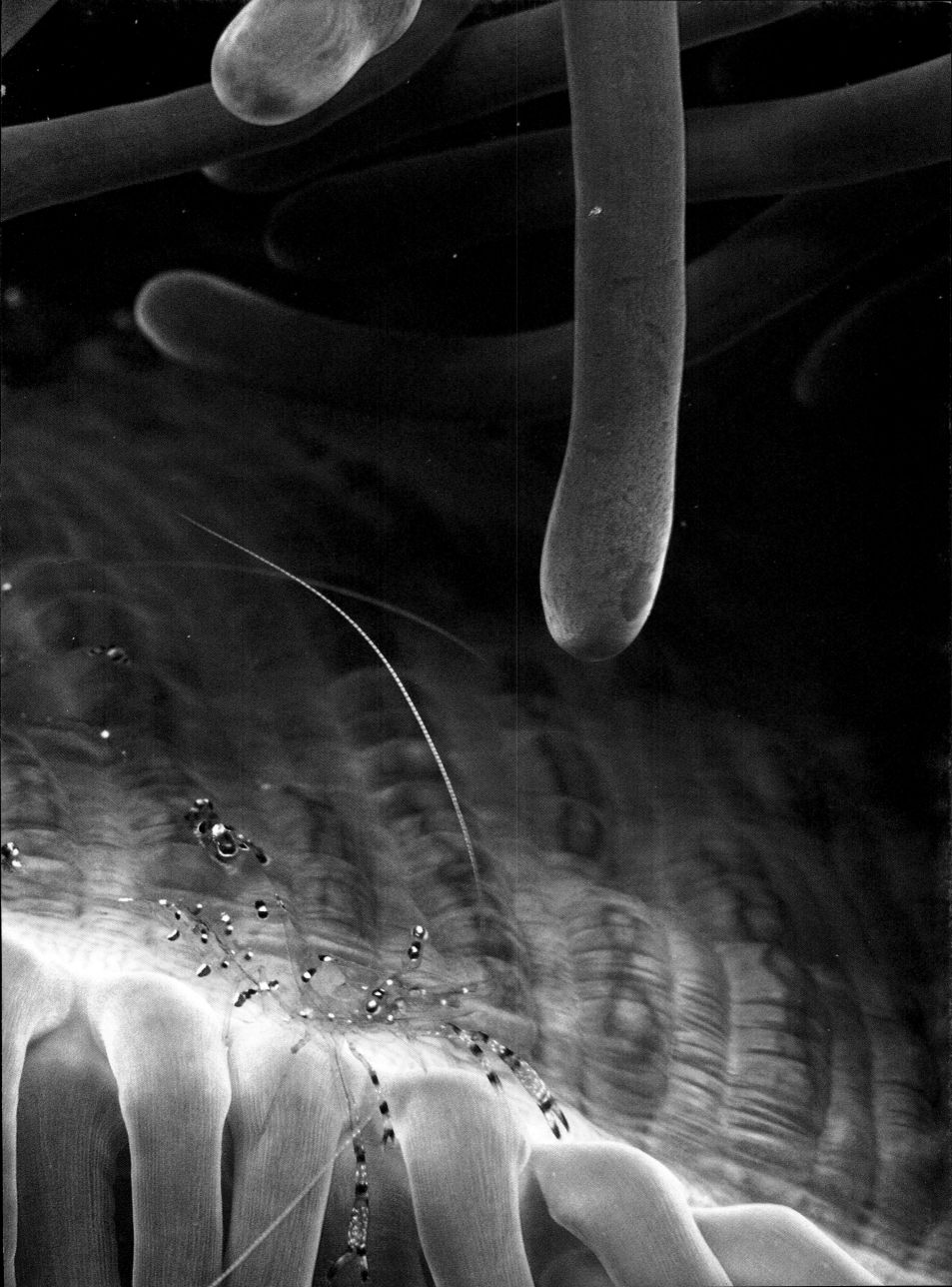

Preface

When a scientist, a researcher, or a curious naturalist comes across a problem, he asks himself a series of questions. How did this happen? How has a living creature learned to cope with a certain situation? What strategies does a species adopt to survive and evolve?

All these could really be summed up in a single question—"How?" How (first and foremost) has such an enormous variety of species and forms evolved on Earth? In this book we will examine one of the more obvious aspects of the phenomenon of evolution—a topic that is often overlooked. We are referring to the subject of *color*.

Let us begin by stating a simple fact. Unlike other mammals, human beings (and members of the ape family) are eminently visual beings. Our senses of smell and hearing are not on a par with those of many other animals, but our sense of sight is quite highly developed. It is perhaps for this reason, as well as the fact that we are surrounded by color, that we tend to take it for granted and feel that it requires no explanation. Why are plants and animals colored? Why do we find such a wide range of reds, greens, purples, and yellows, for example, and not just a uniform world in black and white or infinite variations of grey? Furthermore, how has color evolved? How is it modified by different species according to their particular needs?

Those colors which could be defined as basic in our world—the green of plants, the brown or grey of mammals' fur or birds' plumage—do not confer any advantage apart from the purely functional aim (for plants) of absorbing sunlight or (for animals) of covering the surface of the skin for protection or for the absorption of heat.

In this book we will be examining the ways in which these basic colors are modified and the reasons for such modifications.

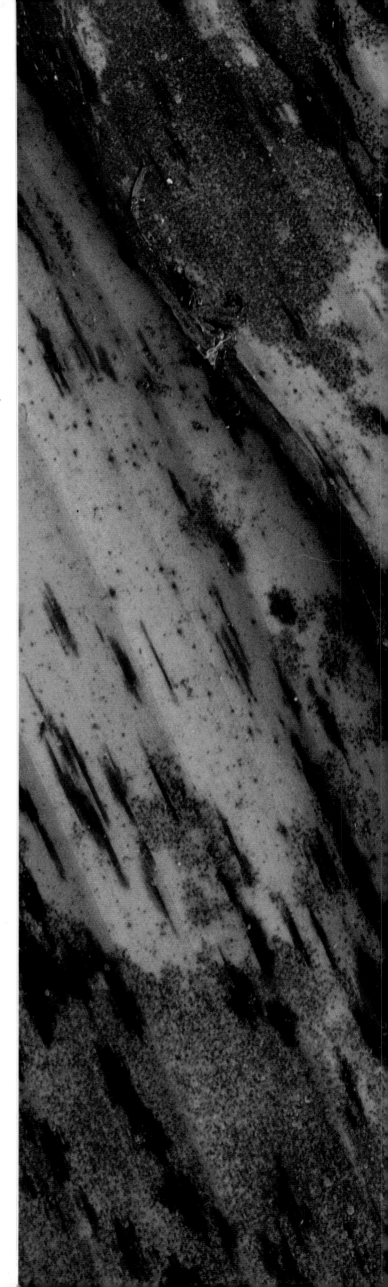

A mimetic animal must not simply adopt the exact color of the substratum on which it rests. Often it is necessary to take on a subtle mixture of contrasting patterns of similar colors, which will appear both strange and surprising to a would-be predator. In the tropical forests throughout the world, there are great numbers of mimetic animals. Among these, butterflies are undoubtedly the most commonly found species. This saturnia in Costa Rica is a perfect example of homeochromatism—similarity to color of environment.

PAGES 16-17
Leaves are among the substrata most commonly imitated by mimetic animals. Some species, like this tettigoniid in the rain forests of Peru, even manage to imitate the parasitic fungi that grow on the leaf.

PAGES 18-19
In direct contrast to cryptic coloration, the phenomenon of warning coloration is present in animals that defend themselves by means of their powerful poisons, as in the case of many tropical amphibians such as this red-eyed tree frog in Costa Rica.

Introduction
How Animals See Color

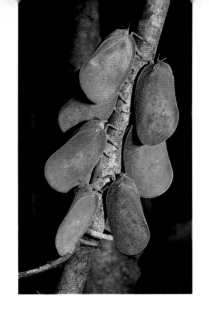

When men began asking themselves what light was and where colors came from, they made some very interesting discoveries. Isaac Newton, for example, thought that light was transmitted in the form of small "packets" (or "particles"). His rival, Christian Huygens, a Dutch physicist, believed that light required some sort of support system such as water (which, when it rises and falls, produces waves) or air (which when it vibrates, transmits sound). This dispute would, with time, prove to be an extremely important one.

Today we know that light, or rather electromagnetic radiation, is transmitted in the form of waves or photons, which brings us back once again to the wave-vs.-particle dispute. The so-called duality of photons is one of the concepts of modern physics, and one which has yet to be fully understood.

What interests us, however, is the fact that the rays are divisible according to length. Those having a wavelength ranging from 400 to 700 nanometers (a nanometer is one billionth of a meter) are perceived by the naked eye as what we call *light*—or rather, *visible light*. Those having a longer wavelength (less energy) are known as *infrared rays*, which transmit heat, and so on until we come to *radio waves*, which have an even greater wavelength. More energetic photons, those having a wavelength of less than 400 nanometers, are the *ultraviolet rays* and the other categories of rays, e.g., X rays and gamma rays. Visible light strikes objects (including living creatures) and is reflected back after having been modified in some way. A part of this light is absorbed, but the remaining, unaltered part is what gives an object its color. Therefore, the yellowish-green plumage of a bird preferentially absorbs all wavelengths except the yellowish-green ones (around 560 nanometers).

The so-called constant nature of color is one of the most universal phenomena among vertebrates. It was once thought to be characteristic only of organisms with complex brain structures, such as man and certain other mammals. However, this phenomenon has also been discovered in goldfish, which certainly do not fit the criteria.

It is interesting to note that the visual systems of animals and the photosynthetic process in plants have a close similarity. Solar energy is trapped, in both the plant and animal kingdoms, by chemical compounds that act as *pigments*. Plants use the rays of the sun to synthesize organic compounds. Animals use these rays to see. Even plants, during photosynthesis, "see"—that is, they perceive the same band of wavelengths as animals. This is because the wavelengths visible to both animals and plants are the most numerous ones that come from the sun. In simple animals, such as flatworms, two small cavities filled with pigment capture light and determine the source from which it comes. The flatworm can distinguish light from dark, but it cannot see a sunset or recognize its mother. As we ascend the zoological scale, the structure of the organs that interpret light signals becomes more and more complex.

An obvious variable that affects the quality of the image is the number of photoreceptors; the process is a little like that which takes place inside a television set. The greater the number of dots being stimulated, the better the image. It therefore follows that the octopus, which has 70,000 photoreceptors, sees a much clearer image than a planaria, which has only 200 photoreceptors. Obtaining a clear, sharp image is another important requirement: the various light rays must be deviated, or refracted, by a lens of some sort, before striking one precise spot at the back of the eye: all this is done in order to obtain an exact image and not just a mixture of light and shadow. Only a few invertebrates possess eyes capable of constructing images. The compound eyes of insects and cephalopoda (octopus and squid, for example) are the most complex and most efficient examples to be found in animals that do not have an internal skeleton.

But for an eye to serve its owner, an image is not enough; it is also necessary to have some sort of nerve circuit that interprets the light and shadow that strike the light-sensitive pigments. That nerve circuit is the brain. We certainly cannot know how insects "see" the world around them, but we can

The phenomenon of group camouflage is not very widespread, perhaps because the behavior patterns of the animals involved must be perfectly coordinated. One of the most classic examples is that of the Phiatids (above), a species of tropical hemiptera found in Madagascar. Their wings are full and brilliantly colored. When they are resting on a tree trunk, as in this photograph, they resemble an inflorescence. In some species, the adults may have two different colors and position themselves on a branch in such a way as to resemble colored flowers surrounded by green bracts. Mimicry is not only used for defensive purposes. It can also be a means of attack. This is particularly so in animals that remain immobile, such as many mantises. In the so-called flower-mantis (right), the triangular appendages along the abdomen disrupt the outline of the body and resemble the flowers on which the mantis rests. The insect is so perfectly camouflaged that insects will actually land on it.

presume that the thousands of ommatidii (the miniscule "facets" which make up the eyes of insects) send thousands of images to the nerve centers and that it is in fact these groups of cells that are responsible for reconstructing the images themselves—by putting them together, rejecting the unnecessary details, and making the necessary distinctions between the image of an object and its surrounding background. All this is done in order to obtain a true color likeness of the exterior world.

The efficacy of every organ of a living creature is constantly tested. Through a creature's dealings with enemies, prey, and danger, the exterior world literally selects those creatures whose organs are strongest and most geared to survival. Natural selection is the ultimate key. In order to serve its purpose, the image "constructed" by the eye-brain apparatus must help an animal to distinguish food and recognize danger. The eyes of insects and cephalopoda have completely different structures, but each has evolved to serve its owner's needs.

This and other facts have often been pointed out by those who are opposed to the theory of natural selection as being examples of perfection, which cannot be explained in terms of purely random evolution. Even after formulating his own theories, Darwin himself was forced to admit that the eye was one of the most difficult structures to explain.

Let us try, therefore, to show how the structure of the eye has been used to explain how natural selection works. The eyes of the cephalopods and those of vertebrates are both efficient and have a very similar structure—except for certain details. So, as a result of the same evolutionary urgency (the need to see), two extremely diverse groups of animals have tackled the problem in a similar, if not identical, manner.

Why should a specially created entity (i.e., one that does not evolve through natural selection) reach the same point by means of completely unconnected evolutionary stages and by means of different structures? Organisms which have developed through evolution must adapt existing structures to new tasks. Renowned biologist François

The spiders of the Tomisiid family are also known as crab spiders because of the lateral gait they adopt when disturbed. They all carry a powerful poison, which they use to paralyze their prey. This species, Misumena varia, *can assume different colors ranging from yellow to white, depending on the flower on which it lands. Such changes are achieved by ejecting a particular substance from the intestine, or drawing it back in when it serves no purpose. Then, often according to the seasons, the spiders move onto other flowers to lie in wait for their prey—the insects that come to pollinate the flowers.*

Jacob compared evolution to a do-it-yourself expert who builds new constructions from pieces of this and that he finds lying around the house, like electrical circuits or plumbing systems which, built up from discarded pieces of different materials, manage to illuminate houses or discharge water efficiently but by means of different pipes and wires.

In all mammals and birds and most other vertebrates, there are two types of receptors. One type perceives forms clearly in black and white, while the other decodes colors but not forms. As well as the different light intensities which reflect the form of an object, evolution has brought about the means by which these same rays are separated into colors. Each elementary color corresponds to a different receptor. In mammals, these color receptors are called *cone cells* and contain various pigments which are sensitive to three basic colors (blue, green, and red). Obviously, since other colors exist as well as these three basic ones, the eyes are able to distinguish thousands of different tones and shades as combinations of these colors in varied brightnesses and hues.

In animals belonging to the higher orders, the eyes have other functions which are equally as important as the function of sight. Eyes can become dangerous targets or fascinating attractions to the opposite sex. Eyes that protrude from the body, like those of some fish living on coral reefs, are cleverly disguised by stripes and other patterns, which confuse predators. Some amphibians use their eyes to threaten would-be predators; the red irises of the South American tree frog warn its enemies: "Stay away from me—I'm poisonous."

Among the cold-blooded animals (in actual fact, not really cold-blooded at all, but in part dependent on the temperature of their environment), the prize for the strangest eyes must go to the chameleon. Their eyes function independently of one another; each sees a different image. What the scientists have asked themselves is "How can the reptile's brain integrate two worlds that cannot be completely superimposed one upon the other?"

Just before the moment of attack, the slow lizard turns both eyes upon its victim. Only then do the two images join to form a single image, and the chameleon obtains the binocular (two-eyed) vision necessary for capturing its prey with great precision. In appearance, the brilliantly colored irises of another reptile, the gecko, (the proverbial "cold reptilian eyes") are similar to the bright red eyes of the eagle-owl, which, however, has a level of keen-sightedness far superior to that of most other members of the animal kingdom.

Birds are visual creatures and extract most of their information about the environment around them by means of light: our world, therefore, is paradoxically more similar to theirs than to that of some of our mammalian cousins. Birds can see in color and many of them, like us, have binocular vision. Their brains are structured in such a way as to interpret a world of colors.

Pigmentation

An animal or a plant is made up of different colors for varied and complex reasons. In the majority of cases, the most immediate explanations involve compounds which biologists call *pigments*.

These are molecules, usually very complex in structure, which serve a variety of purposes in the body. The ones called *melanin* give scales, hair, fur, and feathers their color. Melanin is derived from the amino acid *tyrosine* as a by-product of digestion. The *carotenoids* are yellow, red, and orange: they are synthesized by plants; animals can can only take advantage of these pigments if they feed on plant life.

All these compounds can be incorporated in small corpuscles inside the cells or even occupy the whole cell area. They are stored in the body and can appear or disappear at different moments according to whether an animal needs them. In plants, these pigments are of fundamental importance; the chlorophyll present in most plants collects the light energy that the plant needs to live. These colors could be defined as naturally occurring "real" colors.

Many of the basic colors we see in nature result from the selective absorption already described: in plants, chlorophyll molecules selectively re-reflect photons whose wavelengths make them appear green. Other colors, however, are produced by light which is "rejected" by an object so as to produce truly extraordinary and ephemeral effects; these are known as the "structured" (or structural) colors. In the same way as light strikes a prism or a drop of oil on water divides into different colors, so too does a ray of light striking a bird's feather of a particular structure become iridescent and appear to change color, according to the angle from which it is observed. The bright colors of birds' feathers and fish scales are, in fact, produced by the diffraction of light.

Other structural colors are caused by *interference* between two or more colors, or from a phenomenon called the *Tyndall effect*, which is created as a result of minute particles that diffuse light rays. A stratum of diffusion which brings about the Tyndall effect is to be found in the wings of many birds, in chameleons, in frogs, in snakes, and in lizards. The colors produced by interference are to be widely found in the fanciful plumage of birds.

Light is reflected by a layer of keratin in the feathers and also by the underlying stratum. When this happens, the rays of light reflected by one layer become interfused with those reflected by the other, creating shimmering colors—again, like those of a thin layer of oil on water. Colors like those on the neck feathers of mallards and starlings and even the beautiful plumage of hummingbirds take on a metallic sheen as a result of this effect. Interference is also responsible for giving butterflies, fish, and many other birds a wide range of colors, particularly in tropical regions of the world, where color often signifies the difference between

being seen (and consequently finding a partner) and being ignored.

Some animals have skin that lacks pigment entirely. Those who have aquariums will no doubt be familiar with the *kryptopterus bichirrhus*, which, against a neutral background, seems almost to disappear. Another animal with a transparent skin is the glass tree frog, which, when resting on a leaf in its native habitat (i.e., the tropical forests of South America) takes on the green color of the foliage beneath it. In these creatures, the pigments have disappeared for quite definite reasons. But we could also ask ourselves, "Why do pigments exist at all?"

A male frigate-bird in the Galápagos inflates the scarlet pouch beneath its bill to attract a female. The males often gather in groups upon the mangrove, where they will later build their nests, and engage in this behavior so that the females flying overhead can choose their partners. The scarlet color of the pouch contrasts sharply with the black feathers on the upper part of the bird's body, which have a metallic sheen. The throat pouch is red as a result of carotenoids, and the dark plumage is iridescent as a result of a phenomenon known as interference, in which the light rays from the feathers are reflected in different ways.

Chlorophyll

Without one particular pigment in plants, we would not be here asking ourselves about the significance of color. The pigment we are referring to is *chlorophyll*. With this compound, light and color have had a dramatic impact on life on earth, making significant and irreversible changes to the ecology of our planet. Chlorophyll enables green plants (and some forms of bacteria) to use solar energy in a beneficial way. In fact, this pigment has the extraordinary capacity of being able to absorb some of the different wavelengths of light and convert this solar energy into molecular energy—in particular, into sugar molecules. A waste product of this process, which begins with carbon dioxide and water, is oxygen, which we breathe every minute of the day.

Without the first microscopic green plants, the Earth would not have taken on the beautiful greenish-blue color seen by the astronauts from thousands of miles away in space. More importantly, animals that use oxygen to breathe could not have developed at all. The process of evolution that began with these animals and then led to the appearance on Earth of the first vertebrates, fish, and finally, mammals, has been a relatively short one. Chlorophyll is, therefore, a pigment without which we could never have evolved and one that is essential to our survival.

Because chlorophyll absorbs, in particular, the wavelengths corresponding to violet and red and re-emits green and yellow, we perceive the leaves of plants as having a green color. Some marine algae that live in areas where water has almost completely obliterated light have had to develop special pigments to absorb that small amount of light which manages to penetrate the ocean depths. This is also the reason why some algae is red, these being the algae that absorb the green and yellow rays. But with a few exceptions, all plants are green in color, or, to be more precise, their most important structures are green in color.

The shades and tones of green may vary considerably, and there might also be red veins or yellow markings present, but the dominant color

of leaves is green. In autumn, chlorophyll is "withdrawn" from inside the leaves; however, the red and yellow pigments remain, and these give the leaves their autumnal colors. Green plants are a classic example of what biologists define as the "physical limits of evolution." Plants are green because they contain chlorophyll; if a plant were to lose its green color, it would lose the main reason for its existence—photosynthesis.

However, if certain structures are not necessary for the day-to-day life of a plant and if the leaves satisfy its nutritional needs, those structures need not be green. Like all living creatures, plants have organs that are specially designed for repro-

duction; flowers are organs of reproduction. Even though botanists refer to flowers as "modified leaves," these are rarely green in color; they stand out from the main structure of the plant in order to attract insects which will help them to exchange genes with other plants. This last point comes under the the use of color as a signal in relationships between species, which we will address in the next few chapters.

In the majority of cases, snakes are predators that lie in wait for their prey; their mimetic abilities reach high levels of perfection. Shown in the photograph on the facing page is a Palm viper from the tropical rainforests of Costa Rica. Another snake belonging to the same species is the

"Eyebrow viper," pictured here, so-called because of the strange browlike structures above its eyes. The difference in color between these two species exists because they come from different environments.

PAGES 32-33
The larvae of the Emperor moth feed on heather. Many caterpillars are defenseless and try to escape their predators by means of their colors.

The Functions of Color

As far as animals are concerned, it might seem that colors only serve the purpose of providing camouflage. However, even in animals, a primary purpose of color is to provide protection from the sun. Our own skin pigment, melanin, is indispensible for survival, especially in areas where there is a large amount of ultraviolet light.

Another purpose is the absorption of varying degrees of heat; in fact, an animal's body temperature is governed by the color of its skin. A dark skin absorbs more heat than a light one. However, the light coloring of most northern animals is also a result of the more complex structure of their fur, or coat, which absorbs the heat of the sun and then transmits it to the body.

In almost all vertebrates, the color of the body is not governed by the skin itself, but by the structures that cover it, such as the scales of fish and reptiles, the feathers of birds, and the fur or hair of mammals. Among the vertebrates, the most brightly colored animals are amphibians and birds. In the former group, however, it is the skin itself which is colored, while the latter possess pigments, such as melanin and certain carotenoids known as the *xantophylles*.

In mammals, the fur or hairs, which contain varying degrees of melanin, dictate the color of the animal's coat. That color gives protection against the rays of the sun is clearly demonstrated by the fact that the majority of deep-sea fish and almost all amphibians and fish that live in caverns do not have colored skins. Since light rays do not reach them, they do not need protection.

It cannot be denied that this voyage of discovery into the colorful world of animals and plants has provided us with a complex and fascinating overall picture, and yet the limits and the physical properties of colors are so numerous that we might be running the risk of thinking that the whole process is a little mechanical.

But beyond their function as sunscreens, colors are among the most powerful mechanisms created by evolution for communicating or concealing the presence of a living creature. The life or death of an animal or plant depends upon them, as does its ability to reproduce. The future of a population or an entire species depends upon them.

Predators take advantage of camouflage when lying in wait for their prey. This lion crouching in the tall grass of the Kenyan savannah is an important example of uniform coloration being used to take prey by surprise. The fur coats of other carnivores such as the tiger have contrasting colors which, nonetheless, blend in perfectly with the play of shadow and light within the forest. The markings on two other large cats—the leopard and the cheetah—help to break up the outline of the body.

PAGES 36-37
The polar bear's dense, white fur, combined with layers of fat beneath the skin, protects him from the bitter cold. But the white fur serves another purpose—it serves as camouflage when the bear is hunting.

Color as Camouflage

Anyone strolling through the woods of southern England on a particular day during the 1950s would have happened upon a most unusual sight. A distinguished gentleman was trying, for all he was worth, to catch what looked like insignificant small butterflies. His attempt to capture these butterflies and the conclusions of his research are an example of the contribution that amateurs can make to scientific research.

Dr. Kettlewell had been a physician for fifteen years before changing the course of his life. Having always been an enthusiastic collector of butterflies, he had noticed that on the industrial outskirts of cities such as Birmingham, a particular butterfly known by the Latin name *Biston betularia* was changing its color. The usual form, which, to do justice to its name of "betularia" (beech-colored), was pale in color and lived on the trunks of birch trees, was becoming increasingly rare and was being outnumbered by another form known as "carbonaria" (coal-colored). The latter, which was black as pitch, was becoming more and more widespread; but it could be seen on the trunks of birch trees and would have been naturally conspicuous against that background. However, as a result of increasing atmospheric pollution in the industrial areas, the trees had become completely covered with soot. Therefore, against the now-dark background, this black butterfly managed to elude the predatory birds in the area. On the same tree, however, the pale specimen now stood out like a snowflake in a coal cellar.

Kettlewell, after catching these butterflies, carried out a long and painstaking series of experiments upon them to establish the reasons behind the change from "betularia" to "carbonaria." Obviously, the precise details of the experiments carried out over many years are both numerous and complex, but the conclusions were quite clear. A pale butterfly on a pale background (or a dark butterfly on a dark background) had less possibility of being detected by predators than a creature whose coloring contrasted with its physical background. With these observations, Kettlewell managed to give one of the first demonstrations of the importance of natural selection in the transformation of animal populations. In the process, he also came to understand a great many other things.

First of all, in every population, forms that are different from the "norm" will always exist and continue to be reproduced, regardless of the mechanisms of selection. In addition, the shape and color of an animal's body are determined according to its predators to a greater degree than was thought possible.

Once its main physiological needs are met, an animal or a plant can use the various colorings on the surface of its body to increase its chances of survival. And eluding predators is in many species one of the prime necessities. As we saw with the *Biston betularia*, one solution is to hide or even to always exist in an environment which has similar color to one's own. This does not mean, however, that predators can't take on colors too—that render them invisible to their prey.

The first solution, therefore, is to have a grey, brown, or green color—a color so widespread that it would blend into almost any environment. This is the simplest form of camouflage, a form which, in one way or another, most animals use. An antelope has a beautiful reddish-yellow color, like the dry grass of the savannah, but so does the lion. In the dense forests, dark-colored animals are not rare; the black panther (which is basically a leopard whose fur contains large amounts of melanin) is more widely found in the forests of Asia than in the African savannah. The majority of small desert rodents have the same coloring as the sand around them—and the same color as the rattlesnakes and vipers that hunt them.

There is no doubt that color sometimes results from other factors — the absorption and emission of heat fundamental among them. But, on a parity with function, why exactly *that particular* color? Definitely so as not to be discovered while trying to absorb some of the heat from the sun. These two fundamental requisites of evolution—thermo-regulation and camouflage—which, at times, contrast with one another, have come together in a kind of reconciliation. In this way, a compromising color that serves both to heat and to hide an animal can help it during all the moments of its life.

The function of the coloring which biologists term *cryptic* (i.e., which confuses the animal with its background) is that of making the animal

In the photograph above are two myriapods that have developed almost identical coloration to the plants on which they live. Pictured at right is a stick insect. Stick insects and leaf insects (the Phasimids) camouflage themselves in order to escape their predators. Because of their extremely slow movements and their highly modified body structure, many stick insects have no wings. Those species that do have wings can open them suddenly to frighten would-be predators. Mimetic ability, however, is not confined to adult insects; the eggs are also camouflaged. On the ground, they look very much like vegetable seeds.

"disappear" from the sight of its enemies when they are hunting for prey.

At this point a brief digression into the subject of ecology is necessary. If this were all there was to it, if the biology of the animal and all its strength were concentrated on merely trying to avoid being eaten or exploited by some other species, everything could be quickly resolved; the animal would just move far away from the area frequented by its predators in order to be safe. However, the prey has reasons for wishing to stay in that area where it is hunted.

In the majority of cases, the hunters and the hunted are neighbors for purely alimentary reasons. An herbivore lives where it can graze; the flowers an insect needs sometimes only grow in a particular place. It would seem that this should be the opposite: even if the herbivore moved, predators would follow, since predators live in areas where there is an abundance of animals on which to prey. In any case, the hunter and the hunted live together in the same habitat. One of these must do all it can to avoid being seen, even if, at the same time, it has to feed itself, move about, take shelter, and make itself visible to the opposite sex so that it can reproduce. Without doubt, what experts observe in the behavior and biology of every species are the constant attempts to fulfill these diverse requirements. How can an animal survive its predators without becoming a predator itself?

Very broadly speaking, *communication* refers to an exchange of information between individuals (including members of different species). The ways in which animals communicate hold quite a few surprises.

One distinction to make is between voluntary communication (in which each individual uses colors for protection) and involuntary communication (in which the colors of the prey reveal it to the predator). The prey, for its part, would be quite happy to remain unnoticed. However, it is obvious that the best way of dealing with a predator is to reduce involuntary communication to a minimum. In short, one should do everything possible to make sure that one is not discovered. What goes on between these animals is a kind of evolutionary one-upmanship.

Widely found throughout the world, stick insects (opposite) do not simply use their coloration and body structure to escape their enemies. Some, when disturbed, go into a catatonic state—they appear to be dead and fall

to the ground like broken twigs. Others engage in chemical "warfare" and emit a foul-smelling liquid from their bodies. The photograph above shows an oak inchworm caterpillar that perfectly resembles a twig. All the parts of a tree are imitated by insects—by caterpillars and stick insects in particular. Measuring worms are particularly gifted at this type of camouflage.

Mantises are among the most common predators in the forests and in the savannahs of the hottest regions of the globe. The entire body structure of a mantis has developed as a result of predation. It has a triangular head with two enormous eyes, a long prothorax, and armlike forelegs that grasp and clutch their prey, which they prefer to eat alive.

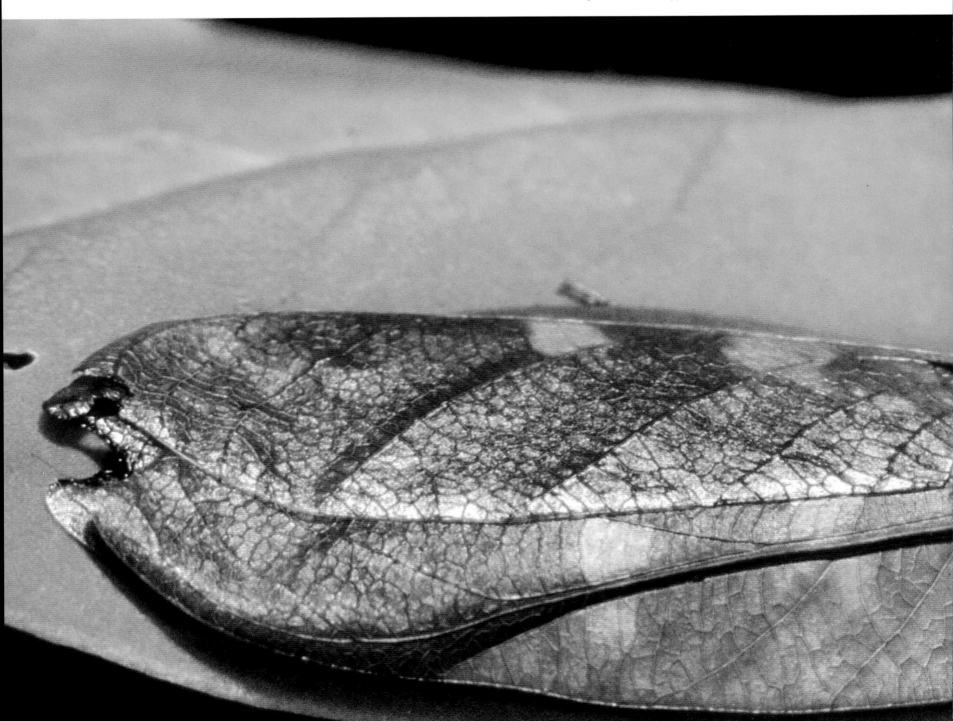

PAGES 44-45
All the stages in the life of a mantis are mimetic. This nymph mantis of the tropical forests of South America is almost identical to the dry leaves of a fern. If it were to rest on a green leaf, it would be easy to spot. Most species take great care to find suitable environments in which to lie in wait for their prey.

PAGES 46-47
As opposed to the majority of other mantises, whose bodies have a long and narrow first segment or prothorax, this Costa Rican species has modified the segment so that it resembles the leaves on which it lives.

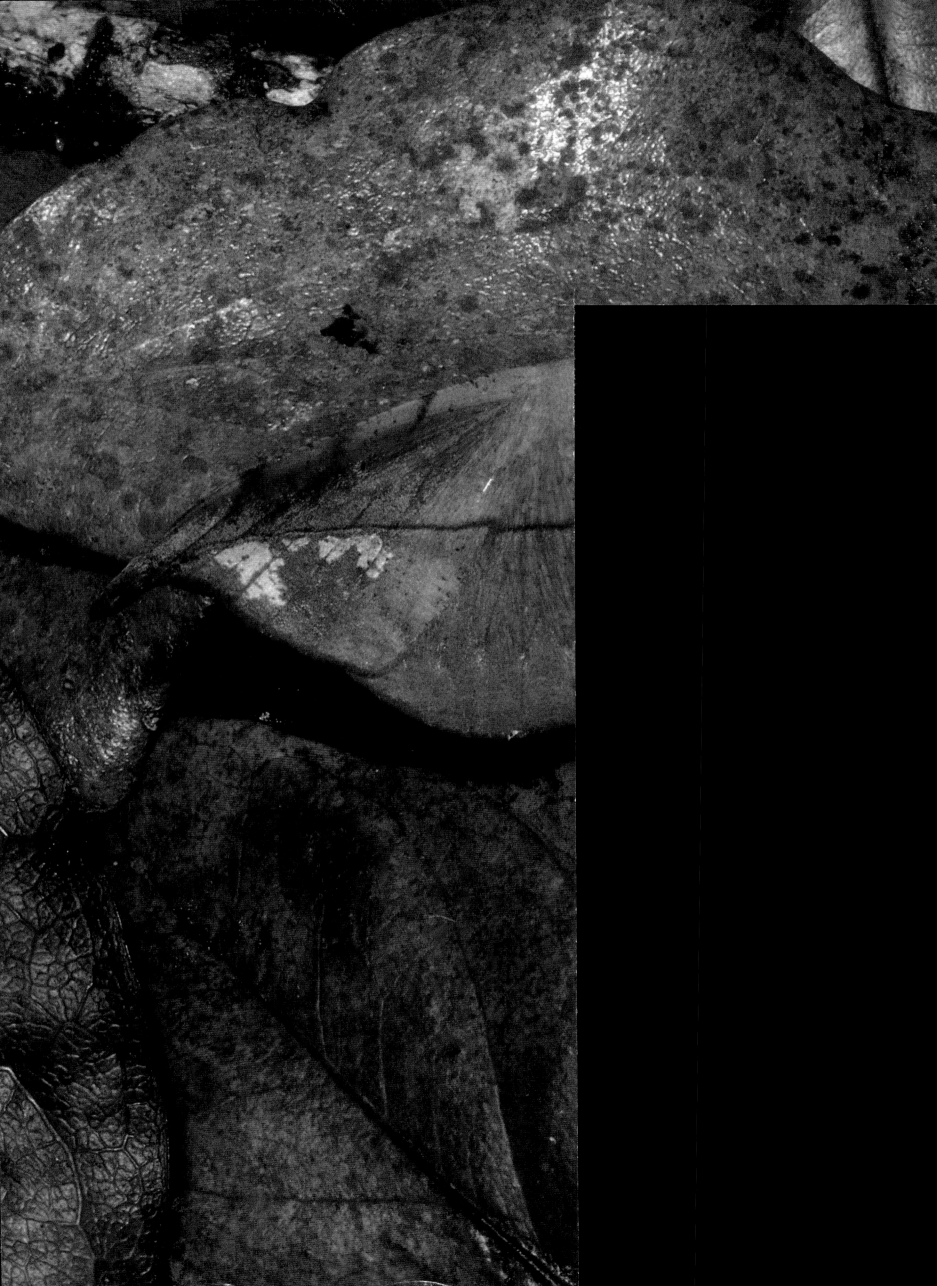

PAGES 82-83
With its wings spread, this moth appears to be one of many fallen leaves on the floor of a forest in Costa Rica.

PAGES 84-85
A small, leaf-shaped frog blends in with the substratum of a forest in Malaysia. The camera has captured the parts of its body the frog is attempting to hide. The frog is best camouflaged when viewed from above—that is, from the angle at which the predator is most likely to view it.

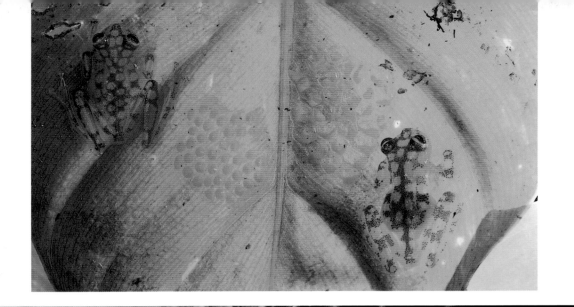

Leaves are among the substrata insects most frequently imitate, as they are found in such great abundance in nature. Both the green leaves still on the branches and those that have fallen provide endless possibilities for imitation. Most leaves have the same structure: a central vein and two more or less symmetrical halves. Among the common insects of the forest, those that seem to have mastered this mimetic art are the Tettigoniae—the grasshoppers. Since a grasshopper's presence could be betrayed by its long legs, these sometimes take on a different color from the rest of the body, as can be seen in the photographs on this page. The legs of some grasshoppers, rather than taking on a coloration that contrasts with the body, attempt to "disappear," as is the case with the grasshopper seen here. The legs of this insect are, in fact, almost transparent. In addition, in some species, the ability to imitate leaves covered with parasitic fungi is extraordinary.

PAGES 50-51
A nymph grasshopper imitating a leaf. In these insects, the wings develop externally and gradually through a series of moltings. The thorn-like spurs on the back legs are yet another means of defense.

Cryptic Coloration

Let us follow, step by step, a predator in search of its prey. A tufted titmouse leaves its nest in the morning intent on finding food for its young. Along with it goes its survival, or rather, the survival of its genes, which are present in the young chicks who wait behind in the nest. While we follow the titmouse, let us also trace the various devices that every species must employ in order to avoid being eaten.

The hunting strategies of a predator are based on a "mental picture of the prey" based on both instinct and experience. This picture consists of a particular arrangement of forms and especially of *colors*—an ideal model of prey which helps an animal to find what it is looking for much more quickly. We, too, when looking for something, unconsciously create a mental picture of what it is we want, and if the object does not correspond exactly to our own preconceived picture, then finding it becomes more difficult.

It is obvious, however, that this "picture"—this arrangement of special colors—need not describe a single sort of prey; with few exceptions, predators will hunt more than one type of prey. The titmouse, on leaving her nest, will be "thinking" about the full green color of the caterpillars hiding under the leaves or the bright red color of the small arachnid parasites on the trees, much as a fox would be "tuned in" to the tawny brown color of the fieldmouse or the grey fur of the hare.

One of the most elementary tricks of survival in nature is blending into the background using colors to avoid becoming prey. The best way an animal has of doing this is to transform itself into one of the features of the landscape—one which does not interest the predator. On being pursued by the titmouse, the butterflies set free in the woods near Birmingham by the kind Dr. Kettlewell instinctively landed where their color would keep them hidden from the bird—on the greyish-white trunks of beech trees. Their coloring, which was so similar to the color of the lichens, provided them with perfect camouflage. And so, with a bit of luck, they managed to escape the bird's attention.

Sometimes the trick of cryptic colors works. At other times, the titmouse will somehow become aware of the butterflies and devour them. For one of the two contenders locked in the battle

A spider of the Tomasiid family lurks on an orchid in an English garden. These spiders have two pairs of extremely sturdy back legs. One might wonder how creatures such as this recognize the substratum best suited to their powers of imitation. Carefully controlled experiments have revealed that, in some species of grasshoppers at least, stimuli from the environment itself guide the animal. The insect stops moving only when the color and configuration of the plant are those best suited to hide it.

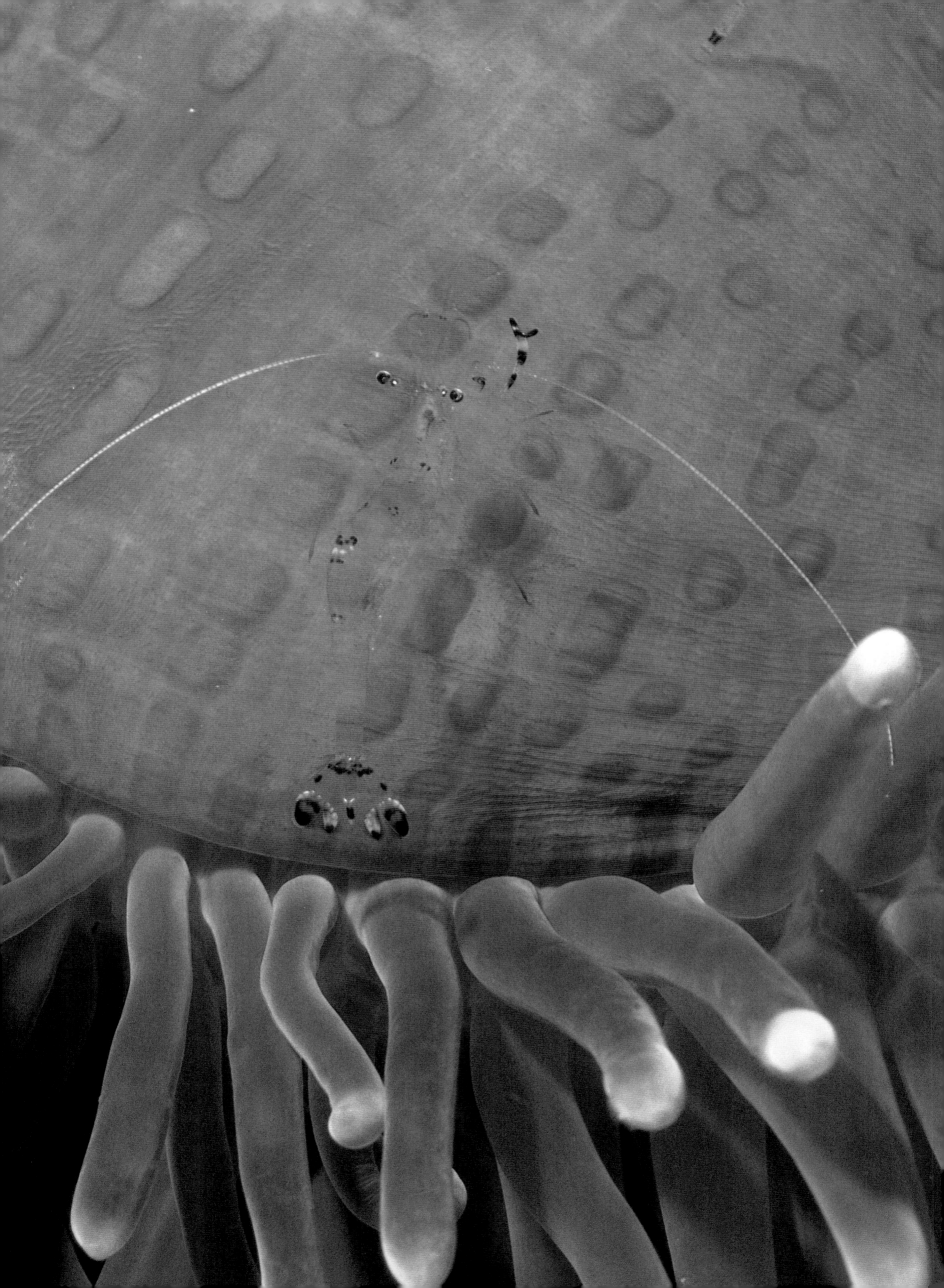

for survival, one tiny detail that gives away its presence can prove fatal. But when can a creature be defined as a cryptic animal?

An animal that resembles the substratum upon which it lives is *camouflaged*. If, on the other hand, a species has the same shape and color as an object in its native environment, it is more correct to call this phenomenon *mimetic*. To be even more precise, if an animal, or even only a part of an animal's body, resembles a random portion of the substratum upon which it rests, very often a predator will not realize that it has a meal within reach.

Theoretical speculation about these details has reached extremely complex and sophisticated levels. These have involved mathematical analyses of the "random" concept—inherent qualities of the colors which must or must not be present in order for an animal to be labelled "cryptic," and even curious experiments carried out to try to establish how each animal perceives the world around him. How does a predator see the substratum? How does its brain perceive colors and distinguish between the prey and its surrounding background? Does the prey itself "know" that the place in which it is located actually protects it from attack? And—this is an important question, one to which we will be returning later—how did natural selection manage to generate forms and colors which are so extraordinarily similar to inanimate objects?

In general, *evolution has seen to it that a cryptic animal resembles something which holds no interest for the predator*. Natural selection has imaginatively transformed the colors of hundreds of species to resemble leaves, rocks, sticks, and even excrement. In every environment there are innumerable ways to be mimetic or camouflaged, since there are as many different substrata an animal can resemble in order to avoid capture. Perhaps the titmouse steps over or even tramples on scores of small insects unwittingly before actually capturing its prey.

As soon as you enter the forest, you notice that in the undergrowth, the carpet of leaves and twigs is a complex mosaic of random fragments. Different species of trees add different colors to the mosaic. The sun penetrates the forest canopy at strange angles, and the light is broken up into hundreds of colors and shapes.

The moths that inhabit such a rich and varied environment may resemble fragments of dried leaves or the edges of leaves (including those parts that are in shadow), complete with veins. Resembling the thousands of dry twigs that cover the forests of the world is the "specialty" of insects of the Phasmidae family (stick insects).

While you are walking along, a large bird flies up into the air in front of you and startles you; it is a nighthawk whose camouflage has been so effective that you've nearly stepped on it. Up and down the tree trunks, small tree-creepers look like pieces of bark. In the fields, the bright green grass provides background cover for large grasshoppers, which can only be distinguished from the surrounding leaves when they fly away. The important thing for these animals is that they do not contain in the design or the coloring of their bodies any particular details that might attract the attention of predators.

In other environments, other circumstances have been responsible for triggering the processes of natural selection. The predators in these environments are different from our titmouse, and their ways of perceiving the presence of other animals is also different. In the desert, and on beaches, some grasshoppers bear an incredible resemblance to pebbles and small stones. The most extraordinary examples of cryptic camouflage can be found in the Sargasso Sea; here, a complex community of creatures lives in a unique environment among the brown, floating seaweed that covers hundreds of square miles of coastline. The sea dragon of the Sargasso Sea and the frog-fish have adopted the shape of seaweed and have developed long extensions that make it impossible to distinguish between them and the long strands of the plants. And in the Sargassian forests—the "meadows of Poseidon," as they have been called, the needle-fish swim with their heads pointing downward, in perfect imitation of the floating strands of seaweed there.

In the northern Tundra, where no hiding places exist and where the predators are strong, fast, and extremely efficient, small pebbles and a blanket of moss and lichen hide the eggs of the small plover, or "envelop" the female eider as she lays her eggs.

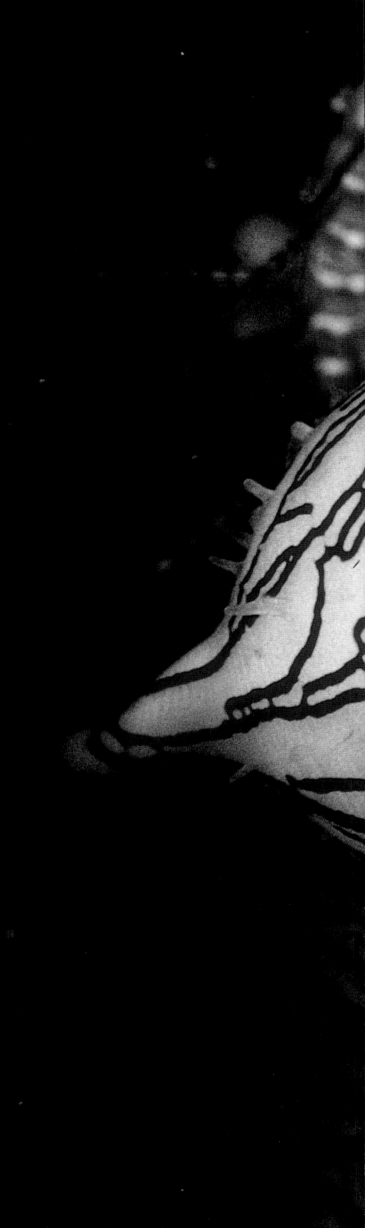

Some species have found it quite easy to imitate the stark whiteness of the snow. Furthermore, white fur or plumage is not "costly" to create and is also useful for storing heat. Hares, snowgrouse, white ermine, and polar bears are all well known for their white fur. Hares and snowgrouse use their white coloring as a means of camouflage in order to avoid being seen by eagles and foxes. White ermine and polar bears use it to avoid detection while preparing to attack their prey. However, if we were to arrive on the mountainside when the snow has melted but springtime has not yet arrived, we might notice a curious feature of the changing seasons. In fact, the alpine hare and the ermine are only white in winter; during the summer their fur changes to either the speckled grey or brown color of their "cousins" living in the flatlands. If winter drags on too long or if the first snow is late in arriving, the poor hare finds itself with the right coloring at the wrong time and stands out all too distinctly.

One of the most curious and useful features of mollusks is their so-called "cloak"—the tissue that covers their bodies and forms a "second shell." As these photographs reveal, the cloak slowly covers the shell and interrupts the chromatic uniformity that could betray the animal's presence. As in many other cases, imitation here depends on the context of the animal's environment. On a variegated surface such as the one on which the mollusk is resting, the brightly colored cloak helps the mollusk to "disappear."

Many animals are able to trigger a series of actions that enable them to imitate the substratum even more accurately when there is even the slightest hint of danger. Often, even though the animal is immobile, these actions require a great deal of energy; the photographs on these pages give a clear example of progressive imitation. In the photograph on the left, the mollusk is still relatively visible against a background of coral. In the photograph on the right, the protuberances the animal has developed on its body have managed to camouflage it perfectly within a matter of seconds. Many species (marine ones in particular) abandon their mimetic skins and only leave their hiding places at night, when their would-be predators rely on senses other than sight to locate their prey.

These four species of fish are contrasting examples of the use of color. The two members of the Scorpeniid family (below, left), have extremely intricate, striking colors. These fish are poisonous and have very harmful stings; their colors are used to warn their enemies. The frog-fish (above) and the crocodilefish (below) both use their colors to hide from predators. In fact, they are predatory fish with limited swimming capabilities; they lie in wait on the sea bed for prey.

PAGES 62-63
The members of the Soleidea species, such as soles and rhombuses, are extremely mimetic. Their benthic (i.e., bottom-dwelling) nature forces them to imitate the substratum as much as possible.

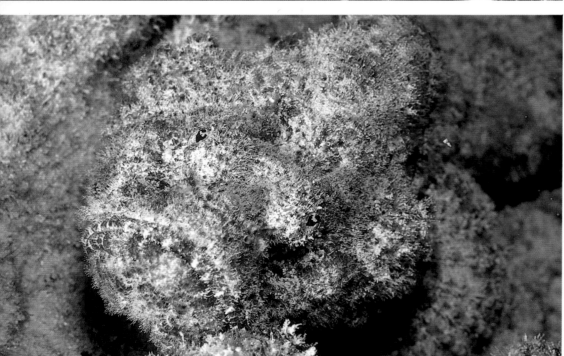

The members of the
Sinanceidea species are fish
with relatively small and
compact bodies. They have
no scales, and instead their
skins are covered with wart-
like protuberances, which
perfectly camouflage the fish
as they rest upon the sea bed
waiting for prey. These protuberances break up the outline of the body, which often
resembles a rock. Often these
fish are able to camouflage
themselves even more by bur-
rowing into the sand. As in
some other species of benthic
fish, the Sinanceidea have a
gland at the base of their
dorsal fin which contains a
powerful poison.

PAGES 66-67
Some gasteropoda such as
octopuses have bare skins
and must therefore protect
themselves through camou-
flage. The two octopuses in
this photograph are using
color to communicate—they
are courting.

Blending into the Background

As we have seen, camouflage, like all other facets of animal behavior, depends on the context within which it occurs. This means that it is not simply a question of being in the right place, but the animal also must be there at the right time. Long and complex experiments have shown that when an American bluejay hunts small butterflies of the *Catocola* family, it will find those whose stripes lie perpendicular to the ridges (or striae) of the tree trunk much easier to catch. If, however, the butterflies position themselves so that their stripes run parallel to the striae of the trunk, they run less risk of being captured. The black butterfly *Catocola antinympha* instinctively chooses a dark background on which to land, while its pale counterpart, *Campaea perlata*, lands on a pale background.

Soles have perfected the art of camouflage and change their color according to their background as well; however, this transformation depends on what exactly they are able to see. If their eyes are near a pale background and the rest of their bodies are against a darker one, the whole body will take on a pale coloring.

The female spider-crab preys on bees and lies in wait for them on flowers. When she lands on a white flower, she becomes white, and when she lands on a yellow flower, she changes to yellow. This is made possible by the presence of a pigment in the cuticle, which she can expand or contract according to her needs.

The caterpillars of the hawk-moth (*Sphingidae*) family are a greenish-blue color, like the needles of the pines on which they live. After the caterpillar stage, they become small brown creatures with black-and-white spots; at this stage they move along to the branches bearing the pine needles, and here again they are invisible. However, the pine inch-moth (or looper) always lives on pine needles, where its green body covered with fine white stripes is "invisible."

The sloth, a creature widely found in the South American forests, has a bushy, coarse coat of hair covered with small green algae. These algae provide perfect camouflage against a backdrop of green trees.

Even the positioning of the wings can give many insects away. Some insects, including numerous varieties of the butterfly (such as the admiral species), have a brightly colored upper body and only the lower body is mimetic. If they open their wings out by just a tiny degree, they are detected immediately by the birds lying in wait for them.

Many small caterpillars take on the appearance of small twigs; they lie on the branches of trees at the same angle at which the twigs themselves grow. The Phasmidae are even mimetic during certain stages of life when most animals are well protected. In fact, their eggs resemble vegetable seeds, and as soon as they are laid, they fall to the ground in the same manner as plant seeds do. On a background of lichen, the Venezuelan *Markia hystrix* is practically invisible. Being a nocturnal insect, it is only found on lichens during the day. Many people wonder how it is able to choose the correct substratum on which to settle. The answer is this: the insect is continually stimulated to move when it finds the wrong background and only stops moving when it reaches the correct one.

This is not to say that a color which provides camouflage cannot also be a bright, vivid one; at close quarters, zebras are anything but mimetic, but seen from a distance among the thick acacia woods where they take refuge, their stripes render them almost completely invisible to predators. The scales of one of the most beautiful fish in existence seem like a palette of dazzling colors within the confines of an aquarium, but they provide perfect camouflage in the fish's natural habitat on the barrier reef. If the spots, stripes, and triangles are even smaller than the smallest details visible to predators, the whole body blends in with the surrounding environment.

However, to be perfectly camouflaged, the animals must be more than just identical to the substratum—they must remain absolutely immobile. Therefore, camouflage is also a question of *behavior*. Obviously this cannot be effectively communicated in these photographs, but inside a forest, for example, only great attention to even the most minute of movements will disclose the presence of small camouflaged creatures. When a stick insect sits immobile on a branch, it is almost impossible to notice, but if it makes even a slight movement, the predator's keen system of perception is immediately activated and the insect is in danger. The same dilemma—whether to move or

Two predatory felines—a tiger (above) and a cheetah (right)—hide in the grass and watch for prey. Perfect camouflage like that found in fish and insects is not very common in mammals, perhaps because most of them are extremely mobile creatures and inhabit different environments, which makes it useless to try to resemble any one substratum. However, a certain type of camouflage known as behavioral camouflage is very common, in which both predator and prey adopt specific postures and try to be in the right environment at the right time—for example, during certain times of the day, when sun and shade combine to create a dappled light.

Within a single environment, it is possible to find predators with different coats who therefore use different tricks to camouflage themselves. The tawny fur of the lion might seem to be the ideal color. In fact, in the dry grass of the savannah, a lioness out hunting (top left) is almost invisible. But the cheetah (left) is not any easier to distinguish in the tall grass; this is because hundreds of small spots break up the lines of the animal's body. Furthermore, these two species hunt at different times of day: the lion hunts at dawn and at dusk, the cheetah during daylight hours. The colors of the lion tend to blend with the environment. The black spots of the cheetah are emphasized by the bright sunlight, increasing the mottled or "broken" effect of the body's outline.

remain still—has to be faced by all the caterpillars of butterflies, which imitate twigs, pieces of bark, and even the droppings of birds.

However, prey can take advantage of a "defect" in perception which is common to the majority of predators. This defect involves the fact that for most carnivores, anything that doesn't move doesn't exist. Therefore, most mimetic animals move only at night, and moths, nighthawks, and other nocturnal birds of prey search for food when their own predators are not around. Some mantises, like those of the *Acanthops* family, resemble dry leaves, and, like chameleons and stick insects, they move slowly, oscillating from left to right like leaves blowing in the wind. One of the most extraordinary examples of aggressive mimicry is that of the mantis *Hymenopus coronatus*, which manages to remain so immobile and looks so very like the flowers of a Malayan orchid that insects will actually land on it to suck in the nectar from its "flowers." Another mantis, the so-called "diabolical idol," actually pretends to be a flower and waits for prey, with its brightly colored anterior legs spread out ready for attack.

Two features that give away the presence of an animal are its shadow and its outline. While some creatures who live in the woods can get around the problem by remaining in areas devoid of sunlight, others must venture out into the sunny areas to hunt and find food. These animals avoid creating shadows by flattening themselves out against the tree trunks. One such creature is a mantis known as *Theopompella westwood*, which spreads its wings in order to conceal its own shadow. In the same way, the gecko of Madagascar (*Uroplatus fimbriatus*) flattens its body against the trunk and "disappears," and the numerous appendages along the sides of its body also help it to camouflage itself. The Australian gecko is clear proof of the powerful effect that evolution has had on methods of camouflage. The *Phyllurus cornutus* has the same shape, coloring, and appendages to break up its outline as the gecko of Madagascar.

Others combat shadow by means of the much more widely diffused phenomenon of *countershadowing*. In a body illuminated from above, the palest part should be the upper side and the darkest part the surface underneath. However, if

an animal's skin is paler on the underside, the animal often takes on a two-dimensional appearance when viewed from above, and its actual shape cannot be correctly judged.

In the ocean, countershadowing is an extremely effective technique: a shoal of fish (sardines or herring, for example), is pale on the underside and dark on the top. Observing them from below, one can only distinguish a patch of color that imitates the illuminated surface of the water. When observed from above, the upper parts of the fish resemble the dark depths of the sea.

A dull, greyish-brown color dominates the various forms of life found in the arid plains of Africa. The chicks of plovers and partridges are light brown and therefore blend in perfectly with the color of the nests. However, if the chicks pop up, their shadows betray their position. Therefore, when the parent birds raise the alarm, the chicks crouch quietly in order to hide their own shadows. The same trick is adopted by ostrich chicks. The ostrich's attempts at hiding have given rise to the generally held belief that the bird hides its head in the sand.

Sharper and darker lines help to blur the outline of an animal. For a predator, this might create the strange effect of seeing only half its prey, and even at that, what is observed does not correspond to the mental picture. Antelopes take advantage of countershadowing and so-called "destructive coloration" (lines and patches of color that blur the outlines of animals) to escape from its predators. The animal's back is darkly colored and its abdomen is pale. In the sultry heat of Africa an impala, when seen from a distance, looks like a mere patch of color on the horizon. Moreover, the abdomen and back are separated by a sharp black stripe, which adds to the effect of its being divided into two parts.

The outline of an animal can also be blurred by vertical stripes that imitate the long, thin shadows of the forest or other well-defined geometric structures. To us, the plumage of some forest hawks (such as the crested eagle of Guyana and the harpy eagle) appears both striking and highly colorful. Even so, they are not easy to spot. The hawk's prey receives no warning because the bird's long crest distorts its outline. The skin of the gabon

viper might also seem to be highly conspicuous, but its prey is fooled by the triangles that cover the snake's body. Many caterpillars of European butterflies simply become features of the landscape, their colored stripes serving to distort the substratum. The larvae of the Satirids, which feed on grass, have vertical stripes, while the larvae of the Pierids, which lie on the leaves of bushes, have horizontal ones. Diagonal lines resembling the veins of leaves are commonly found on the larvae of moths.

Should the need to hide the entire body arise, many species resort to extremely complex and sophisticated behavior patterns. One could almost call these the "extra little tricks" that could mean the difference between life and death. Again, these subtle ploys are based on what the predator *expects* to see.

The animal which is preyed upon has, in the majority of cases, a perfectly symmetrical body, with left and right sides being identical. But if (as is the case with moths) the right wing is placed at a slightly different angle from that of the left one, the symmetry is destroyed and the animal either looks as if it's dead or as if it's something entirely different. Sometimes the symmetry is broken by a wing, sometimes by an outstretched leg, sometimes by a different pattern of markings on each side, and even by the abdomen's being curved to one side.

The small, reddish-colored South American monkey, *Sagvinus geoffroy*, an insectivore, only identifies its prey (stick insects and mantises) by the head and legs. When in a resting position, the stick insect *Metriotes diocles* tightly draws in its legs alongside its body.

One example of the extraordinary and unusual defensive positions that can be adopted by a living creature is that of a small moth from New Guinea. It rests underside up on its extended wings.

Another useful technique is that of group camouflage. Small groups of animals, often interrelated, come together to take on the form of some feature of the landscape. Some even manage to mimic groups of flowers with great precision. In the *Ityraea gregorii* (African insects of the *Phiatides* family), some members of the group are red and others are green. When positioned on a small branch, they mimic the arrangement of color on

Nocturnal or twilight birds like the tawny owl (above) or the podargus papuensis (right) rely on their dappled plumage for protection. During the day they stay out in the open, close to the trees in which they live; the cryptic effect is heightened by their immobility. The podargus papuensis is found in Papua New Guinea. The tawny owl inhabits the woods throughout Europe; the species can be one of two colors, both mimetic: one type has a grey coloring, the other, reddish brown. This owl flies silently in the night, in search of prey.

the blossom—red on the outer edge and green nearer the stem, so that they give the appearance of partially-opened buds.

Let us return to marine life. In the ocean, animals and plants have had more time to evolve alongside each other (*co-evolve*, as a biologist would say), and therefore they have adapted in many (and sometimes exceptionally sophisticated) different ways. The sea is a unique environment, with colors and predators that differ from all others. Here the eyes that need to be deceived belong to fish and cephalopods. Undersea creatures have less sophisticated eyes and less complex brains than ours; this is the reason why we sometimes see forms of camouflage in the sea that we do not find deceiving but that are highly effective in deceiving "local" predators.

Let us move in closer to the grey sand on the bottom. This is not just an expanse, devoid of any form of life. A sole that has been disturbed immediately swims away—undetectable until that moment; its color was identical to that of the sand. In the same way, the mottled skin of the octopus is similar to the kaleidoscope of small animals that cover the rocks. In these cases, the marine species have reached an extraordinary degree of mimetic perfection. Unlike the grasshoppers and the butterflies we have seen in the woods, or the hawks and antelopes in the savannahs, octopuses and soles can position themselves wherever they want. Their ability to camouflage themselves is always with them, and they do not need to look for a suitable background. By means of small muscles, an extraordinary series of cells called *chromatophores* expands and contracts, enabling the colors to change according to the animal's will—similar to the chameleons of the African forests or the anolids of Central America—but at a decidedly faster rate. This phenomenon takes a few hours to occur in soles, but only a few seconds in mollusks (octopus and squid).

However, this does not mean that only dull colors, such as light and dark brown or grey, are used in mimicry. What is important is the substratum upon which each animal lives. If their environment is a kaleidoscope of bright colors and luminous sunny spots, birds that might seem extremely conspicuous to us (such as parrots, golden orioles, and macaws) are extremely difficult to see in their

Chameleons are famous for their ability to change the color of their skin. Since their movements are rather slow, these changes do not have to take place quickly, as they do with fish or gastropods, for example. Moreover, against a green background (in the case of the more common species of chameleon), these changes can be quite limited. Changes in skin color can also take place as a result of a change in the animal's mood and therefore act as a form of communication within the species.

In the sea, the laws of camouflage are slightly different from those that apply on land. This is because of the selective absorption of light by the sea water and the different sight mechanisms of predators. Sea water absorbs light from the surface in a particular way: the red rays are the first to disappear, and below a certain depth, only the blue ones remain. This fact explains the greyish-blue color of many fish. Furthermore, since light always comes from above, the bodies of fish that live near the surface or in shallower waters tend to be lighter on top and darker underneath; this "unifies" their color with the underwater environment and helps them to blend in more effectively.

Predators also take advantage of the protection offered by camouflage, only they do so in order to approach prey without being seen. The Arctic fox is unique among European mammals in that there are two forms of this species, each bearing different colors. The most common form is white in winter and

greyish-brown in summer. The other, known as the blue fox, is greyish-blue year-round. This variability appears to be connected with the animal's need to be camouflaged against every substratum. In fact, in the territories it inhabits, the white fox is more commonly found in the inland regions, where the snow covers the entire countryside. The blue fox, on the other hand, is more commonly found along the coasts. Unlike the snowgrouse we saw earlier, the wild swan (above) stays white even during the nesting period.

As with other Alpine animals, the hare has white fur. The changes in fur color are probably governed by external factors, such as the number of hours of daylight each day, and internal factors, such as the amount of hormones in the blood. During the spring, the fur color changes to brown, but only on the animal's back, and the hare resembles the dark rocks that protrude from the melting snow. It is only during the summer that the fur changes to an overall greyish brown. Since the animal's fur has no yellowish or rusty shades like those of the common hare, the Alpine hare is better camouflaged than its relatives who inhabit the lowlands.

Pictured on these pages are three species of snow grouse that live on two continents; they are examples of how camouflage can change according to the seasons. Snow grouse live in environments that undergo seasonal changes; the birds themselves undergo periodic transformations that allow them to blend into the scenery. In winter, for example, their plumage is white, as in the case of the American snowgrouse (right). Only the black beak and eyes could possibly give it away. (Above) An Alaskan snowgrouse in its transitional plumage. The first snows have begun to cover the landscape, but only the bird's breast blends in; the upper part of the body has not yet "converted" to winter white. (Top, right) A Nordic snowgrouse in its summer plumage. The snowgrouse moults three times each year. Because she must protect her nest, the female's summer plumage is less brightly colored than that of the male.

Sometimes they imitate vegetable and plant structures perfectly and therefore come under the category of cryptic coloration. At other times, however, they are so oddly shaped that no model for them can be found.

As in the case of the super-evolved Pseudophillid grasshoppers we discussed earlier, these protruberances might have more than a cryptic function. Some entomologists believe that these structures contain special sensory organs. The fundamental reason for the dual function is that, even if the animal is perfectly blended into the substratum, it has other needs in addition to that of simply being able to disappear.

If a nighthawk or butterfly that resembles a leaf wants to find a mate, it must, in one way or another, let itself be seen by its prospective partners. At that certain moment, however, just as it can be seen by other members of its species, so, too, can it be seen by predators. Camouflage, therefore, must not be complete: quite often, animals are just mimetic enough to reduce being preyed upon to a tolerable degree without compromising other equally fundamental needs.

In the next chapter we will see how animals try to *attract* attention—not just of prospective mates, but also (for very good reasons) the attention of their enemies.

Although tropical forests offer the greatest possibilities for camouflage, numerous mimetic species can also be found in open and less diverse habitats. The grasshoppers of the Tettigoniid species excel at camouflage both in the tropical forests and on the pebbly beaches along the European coasts. A member of the Tetrigiid species (top), which lives in the humid areas of the tropics, feeds on algae and mud that is rich in organic debris. These grasshoppers imitate their greyish-brown environment as closely as possible. The grasshopper pictured at right bears a remarkable resemblance to the rock upon which it is perching in Karoo National Park in South Africa.

Confusing "The Mental Picture"

Every environment contains examples of camouflage; almost all animals are mimetic in some way. But how did these tactics originate? Evolution by natural selection tends to perfect the successful body structures and physiological details. But the extraordinary ability of some grasshoppers to imitate leaves seems almost excessive.

A positive mutation in the prey can be balanced by a positive mutation in the predator. This could consist of the ability to perceive a difference in *shades* of color. If a parent predator becomes aware of the presence of prey, even prey that is camouflaged, it can pass this instinct to its offspring. For each move made by the prey, there is a counter-move made by the predator. No form of protection is ever perfect. Even well-camouflaged prey are caught by predators, because predators persist in looking where they have previously found prey—under the same leaves, for example.

At this point there are two possible directions in which evolution can go. A living creature can either strive even further for perfection or create "sub-populations," i.e., different forms of its own species that would have different colors and shapes. In the first case, the animal's body becomes increasingly similar to inanimate objects, incorporating extremely sophisticated details until perfection is obtained—an excessive degree of perfection, according to some biologists. Some tropical grasshoppers and some butterflies do not just imitate the leaves or the trunk on which they are resting; they even have marks on their wings that are identical to those the lichens make on the leaves themselves, or that resemble small holes and tears made by other herbivorous insects.

What we see as an interesting green grasshopper, for example, is also a miracle of engineering. The upper section of its wings is covered with tiny raised spots that imitate the surface of blades of grass. The grasshoppers that belong to a tropical sub-species known as the Pseudophillins ("false leaves") are so similar to the leaves of the trees on which they live, so incredibly identical in every detail, that experts suspect this to be a case of what they call *excessive evolution*. These particular grasshoppers look like leaves that have been half-eaten by caterpillars: in fact, they even have patterns that imitate the tiny trails small insects make between the two strata of the leaf.

In the end, mimetic animals motivate predators to increase their efforts to catch prey, to use more caution when hunting, to improve their powers of exploration, and, in the final analysis, they force them to create a mental picture of even the most well-hidden prey. Therefore, the advantages of camouflage can be defeated if the mimetic animals (both as a species and as a number of individuals) become too numerous.

Together with specialization, another way of escaping predators is *differentiation*. In this case, experts talk of *polymorphism* (literally, "many forms"). Since the predators' "mental pictures" are usually quite limited in the majority of cases, the prey that a predator seeks are usually the most common ones. If a population of butterflies, for example, divides up into different sub-populations, it creates the impression that each subspecies is rare. The result is not one population of 1,000 individuals, but rather two or three populations of 330-odd individuals each. The predator still comes across these polymorphic butterflies as often as it did before, but thinks that these are rarer animals and not worth his efforts to catch them, since it is better to concentrate on a more common (i.e., more plentiful) food supply. Further, some of them might seem strange, and maybe even dangerous to the predator. If an animal whose coloring is different from what the predator normally sees should happen to fall into its clutches, the predator's reaction will be one of perplexity; what will happen if it attacks this unknown beast with such oddly mixed colors and which it has seen so rarely? Therefore, some species avoid being caught by "lying" to their predators about their rarity.

Males and females (as in many species of butterflies) can have different colors or belong to different age groups (immature and adult), can have completely different skins, and can all be equally difficult to see. Even the common wood snails (Cepaea) have five or six different populations, each one of which inhabits a different environment. In the Memracids, another species of small tropical insect, the protuberances that grow out of the proboscis are always exaggerated.

Since the creation of special colors and other defenses requires a great deal of energy, living creatures are only mimetic when and where it is necessary for them to be so. In the case of this leopard gecko (above) from Pakistan, only the upper body imitates the substratum upon which it might be detected. The abdomen is almost entirely white, because the small reptile is always careful not to reveal that particular part of its body. The Australian leaf gecko (right) usually lives near trees, and closely resembles bark. It rarely displays its pale abdomen. Even the lizard's eyes are well camouflaged.

Camouflage does not always involve the transformation of parts of the body. This hermit crab (below), which lives along the Pacific coast of the United States, is both protected and camouflaged by the shell of a large mollusk. Sponges have begun to grow on the shell and help the crab to wear a piece of the sea bed wherever he goes. Some species of hermit crab even manage to break off sea anemones from the surfaces of rocks, in which case the crab's protection extends beyond the camouflage: sea anemones carry a powerful poison. (Right) A fish belonging to the Antennariid species, photographed in the waters near Hawaii.

PAGES 100-101
Only the tiny eye surrounded by a white-spoked red circle could give away the presence of this Antennariid waiting for prey on a tropical reef. Some of these fish have almost completely lost the ability to swim.

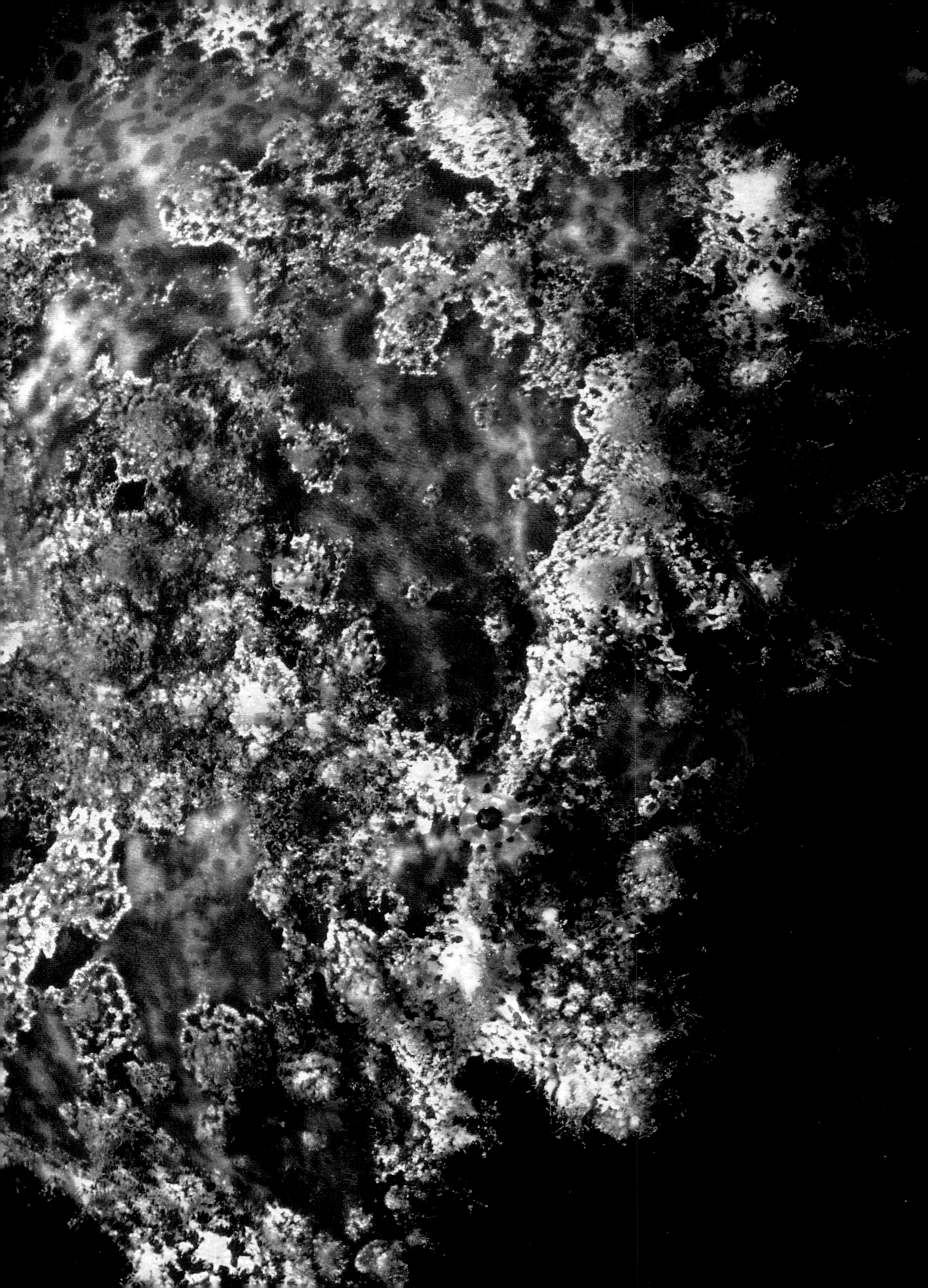

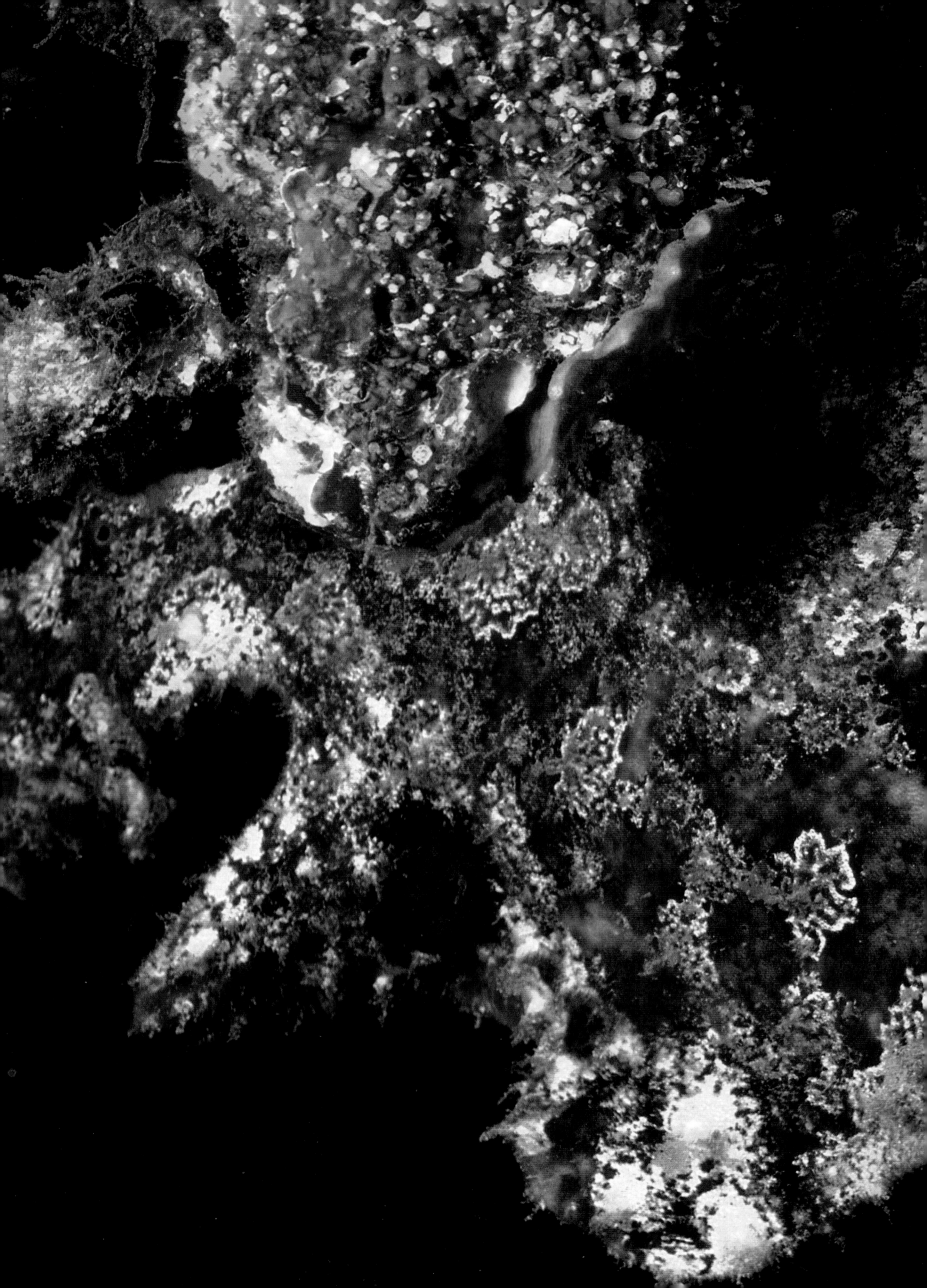

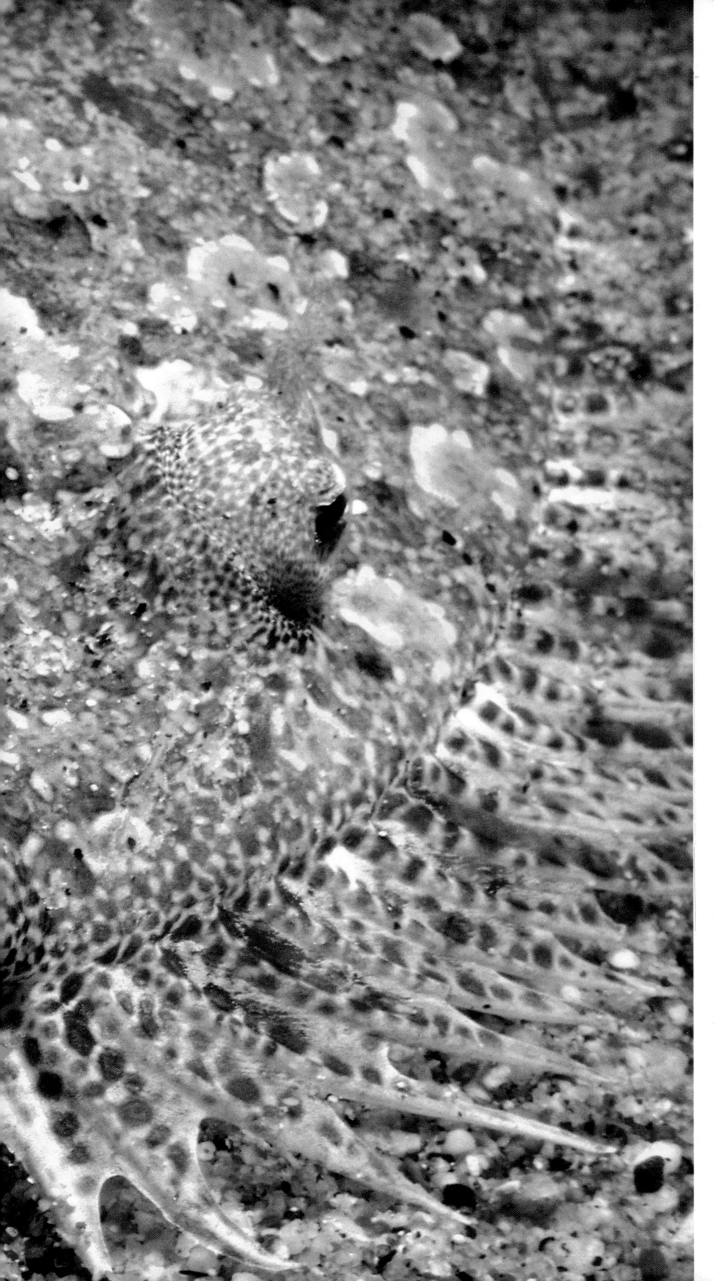

Almost all the members of the Pleuronettiform order (flatfish)— sole and flounder, for example—are benthic (i.e., bottom-dwelling) fish. The development of the flounder is quite remarkable. When it first hatches, the flounder resembles a typical fish. Very soon thereafter, the body becomes flattened and both eyes appear on one side of the head. It is as if the fish is completely restructured in order to spend its life blending into the sea bed waiting for prey. The immobility of the adult flounder calls for it to use camouflage. While some species bury themselves in the sand, others have an extraordinary capacity to change color according to the surface on which they are resting.

PAGES 104-105
Antennariids, such as this one lurking on the sea bed, are immobile most of the time, but can snap up their prey with astonishing speed.

Warning Coloration

Henry Walter Bates was satisfied. After a long day in the forest, he had caught a large number of strikingly colored butterflies. They were exactly what he needed for his collection of Amazonian insects. However, he was also very puzzled. "How is it possible," he asked himself, "that these species continue to exist while being constantly surrounded by hundreds of predators? Birds could very easily catch them, and even an inexperienced entomologist like myself has no trouble catching them." As soon as the butterflies felt threatened, instead of trying to avoid their captors, they starting flying around slowly, as if they wanted to show off their brightly-colored wings.

At first Bates thought that the ease with which he had caught these species was a result of his own steadfast determination. However, if this were true, they would have disappeared a long time ago, not only from the tropical forests, but off the face of the Earth. After pondering the question for a long time, Bates reached a conclusion.

In 1848, at the age of twenty-three, Bates had gone to the Amazon for a long period of entomological research in the company of his good friend Alfred Russell Wallace (who was to return to England before Bates). After eleven years of traveling and collecting species, he decided to return home; in 1862 in the journal of the Linnaean Society of London, his work on butterflies was published under the title, "Contributions to an Insect Fauna of the Amazon Valley."

Just three years earlier, another English biologist (who had himself just returned from a long journey around the world) had published a lengthy study based on his own findings, in which he put forward a revolutionary theory regarding the variety of life forms. This biologist was Charles Robert Darwin, and his theory was put forth in collaboration with the aforementioned Wallace (Bates's friend), who in the meantime had spent a long period of time in Southern Asia. The theory put forward was that of *natural selection*.

Bates immediately realized that his butterflies fit in exactly with the hypothesis put forward by his colleagues. They did not have such bright colors as a result of an act of creation or in order to please the human eye, but rather in order to have a greater chance of survival and reproduce far more quickly.

Furthermore, the colors warned predatory birds that these butterflies were not edible. In short, the butterflies were structured in a particular way—so that their bright colors could easily be *seen*.

If previously, by cryptic camouflage, scientists were referring to a prey's determination to *conceal* its whereabouts from a predator, now, on the contrary, they were seeing something completely different. In fact, these butterflies wanted to *be seen* by any prospective butterfly hunters. The monologue now develops into a dialogue. The predator communicates its intent (willingly or unwillingly) to its prey, but at the same time, the prey communicates its own message that the predator's efforts aren't really worthwhile. And, as there is always the possibility that the hunter will not fully "get the message," it must be communicated loud and clear. The signal must be louder than the "background noise."

However, in the previous chapter, we saw that some butterflies, which seem so distinct to human eyes when displayed in a collection, are decidedly mimetic in their own habitat. The paler and darker patches of light in the forest protect them from the keen eyesight of birds and small reptiles—their most dangerous predators.

But Bates's butterflies are clearly visible even in the almost perennial density of tropical forests. As if to flaunt this seeming handicap, they tend to fly slowly and arduously, even on relatively short journeys. They must possess some quality that protects them, in one way or another, from predators. The unprofitability of eating them must therefore depend on something that the predator has to learn and which it will associate with the colors of these butterflies in the future.

This quality is the poison they carry within them. Their bodies contain a certain quantity of substance (always bitter and sometimes lethal), which they release when they are disturbed or swallowed. These substances are often compounds derived from the plants upon which the butterflies feed. Monarch butterflies lay their eggs on plants belonging to the *Asciepiadacae* family, whose sap contains powerful poisons. The caterpillars feed on these plants and retain the juices as adults; when a bird catches these butterflies, it finds the taste repugnant and avoids them in the future.

The wings of butterflies can assume an amazing variety of forms and colors. It is therefore not surprising that some of the most extraordinary examples of camouflage that exist anywhere in nature are to be found among these insects. However, butterflies protect themselves by other means as well. Most species are very colorful and can be easily spotted by predators. The large Indonesian butterfly (above) has two large spots resembling eyes on its wings, which it uses to startle its enemies. Yet another method is employed by other species, such as the Monarch butterfly (right); in its larval stage, the insect eats plants that have bitter or poisonous juices, storing them in its body. When a bird catches one of these butterflies, it finds the "chemical" repugnant and releases its prey. The colorful wings of the Monarch butterfly make them easily recognizable; therefore, after the first unpleasant experience, birds do not attempt to capture them again.

Experiments have shown how predators learn to recognize and avoid the repugnant species. It is very rare for a predator to know from birth that a particular prey should be avoided. But the "deterrent" is not confined to just a poison within the prey's body. Some species of insects defend themselves through their sting or through other substances which can "shoot out" from their bodies (as in the case of one species of coleoptera, which emits an acidic substance from its abdomen).

Many of the most beautiful fish are also extremely poisonous if swallowed, or, like the scorpion fish (*Pterois*) are powerfully armed. There is no better demonstration of an animal's having complete faith in its weapons than that provided by these slow-moving inhabitants of the warm tropical waters. They are absolutely regal and indifferent in their behavior towards divers and other fish.

Needless to say, many reptiles have extremely efficient weapons, and some tropical amphibians have some of the most powerful poisons in the entire animal kingdom concealed within their bodies. Distinctively colored mammals such as skunks have glands that secrete a strong-smelling and irritating substance. In all of these examples, the distinctive marking or coloration of the animal is known as *warning coloration*.

How have these animals evolved to this stage? The first step in the process occurred when the prey acquired weapons through mutation. Some insects developed an enzyme which enabled them to tolerate the lethal sap in the plant on which they fed and then store the product in their own tissue, making them poisonous to predators.

Many brightly colored marine animals belonging to the coelenterate class (coral polyps, jellyfish, and sea anemones, for example) are simple organisms with simple nervous systems. And yet no one would dream of attacking them. Millions of years ago, they developed such an efficacious and powerful weapon that some believe their evolution came to a halt because their weapons became too lethal. Hidden in their tentacles, thousands of nematocysts (small, urticant cells) catch unwary small fish. The net of tentacles is almost impenetrable. We say "almost" because

one species of nudibranches has found a way of swallowing the nematocysts and then transferring them, still "armed" and lethal, to some parts of their bodies. The poison of snakes comes from modified salivary glands. That of amphibians comes from glands on their skins; in the Dendrobates family of frogs (widely found in South America), these glands developed originally in order to impede the proliferation of colonies of bacteria and fungi that always lie in wait in the humid tropical forests in which

these amphibians live.

Up to this point, all is well; the external defense mechanisms, active even before attack, are extremely effective. However, it is still not clear how the repugnance on the part of the predator has evolved; to gain experience, the predator must, at least once, catch its prey and try to kill or devour it. It is thought that some species of prey are very resistant and have a hard body surface which is difficult to shatter. Monarch butterflies, for example, have very strong wings, and their tiny overlapping scales do not break easily; the poisonous venom, accumulated when they were caterpillars, is concentrated in these scales.

In order for this toxicity to protect an organism, the predator must understand the signal that says this animal is poisonous. Bright, highly visible

These photographs show two different uses of vivid coloration. In the case of the golden toad (below), which is only found in Costa Rica, only the males are brightly colored. They use color to attract females and not as a warning mechanism. Much more threatening to would-be predators is the "poison arrow" frog (opposite). This small amphibian of the Dendrobates species has a poison known as batracotoxin within its skin, which is believed by some to be the strongest known poison in the animal kingdom. The frog's name is derived from the use made of it by certain Amazonian tribes. First they place the frogs on the fire and wait for the skin to secrete the poison. Then they dip their arrows into the milky substance and use them for hunting.

colors can convey such a warning signal. Therefore, here the use of colors is based on completely different principles from those that suited cryptic animals. We no longer have browns, greys, and dull greens but vivid, contrasting colors and shapes. Bates's butterflies were black, red, and yellow—vivid colors, without shadows or lines to blur the outlines. When they flew around, they looked for places and backgrounds on which to land that would make them visible to all.

It is not just by chance that many discoveries about camouflage were made in the last half of the last century. It was during this time that European and American biologists began exploring the tropical forests, a world rich in unknown species. They discovered innumerable species whose evolution and interrelationships had not been explored. Explaining colors became a challenge for biologists who were venturing outside Europe for the first time. After discovering how "the store of weapons" was built up, next came the discovery of the biochemical synthesis of colors. They soon recognized that the colors employed were very often red, yellow, black, and sometimes white—strong, vivid colors that contrast with one another. (Later on, we will also see the importance of contrast between the various *parts* of the skin or plumage.) In the introduction, we saw that the colors involved in camouflage are created by a large quantity of compounds, the most important ones being the carotenoids. It is curious to note how carotenoids, whose production is only "entrusted" to plants (animals cannot synthesize them), are utilized by animals themselves both in their sight systems and also as signals—i.e., to see and to be seen.

Carotenoids are present in various combinations in a great many species. Poisonous sponges, crinoids, toxic sea-cucumbers, and some mollusks are either partly or completely yellow, red, or orange in color as a result of carotenoids, which are also present in the yellow parts of butterflies' wings and in the beaks of birds. The yellow plumage of the goldfinch and the coral wings of the flamingo are all derived from carotenoids. In certain insects, these are emitted by special glands which are yellow or red in color. Curiously, these compounds are usually pale green or even colorless and only take on a bright yellow color in the blood

of poisonous insects. The carotenoids are not only useful as warning coloration; in some insects they are themselves transformed into poisonous compounds, in which case they serve a twofold purpose of being both a weapon and a signal. Formed by the carotenoids or other pigments, these colorations are known as *aposematic*.

There are a great many aposematic animals, in a wide variety of groups. The coelenterates are very often red and orange in color and their colonies cover entire surfaces. But it is among insects that we find the most numerous examples. Perhaps the most widely known examples of harmful insects can be found among the social hymenoptera: wasps, bees, and bumblebees.

The brightly colored cabbage butterfly has toxic compounds the caterpillars extract from the plants on which they feed; one of the most poisonous substances found in nature is contained within the body of the black-and-red Zygena butterfly. When captured by a bird which is not alarmed by the admonitory colors, the Zygena emits from its leg joints a bitter poison, which contains hydrocyanic (or prussic) acid.

If we were to make an excursion to the tropics, the number of butterflies or other vividly colored invertebrates we would see would be impressive. Many carabids or other coleoptera are so conspicuous that they almost seem to be extending an invitation to other animals to attack them; these, too, are repugnant to birds. Yellow and black hymenoptra are equally poisonous. Covered by urticant, poisonous and strong-smelling hairs, these butterfly caterpillars are well armed and highly conspicuous. Many hemipters emit strong-smelling and irritating liquids. Some centipedes have a dangerous sting; their bodies are black with yellow stripes.

Some particularly vivid fish are extremely poisonous; one example is the globefish (or balloonfish), which has black dots along its beautiful yellow body and whose tetradotoxin is lethal, even in small doses. As we have seen, the fish that are least disturbed by the arrival of divers or other enemies are scorpion fish, which are red with distinctive white stripes. The Garibaldi fish contains compounds which are perhaps even more poisonous than tetradotoxin.

As we've seen, the color of an animal can be used in various ways for various purposes. Often, colors that seem to be useful as signals for attracting the female are only adopted at particular moments. Like radio signals that must be concealed from the enemy but received by the ally, these sporadic colors can be used like bright flashes in the dark gloom of the forest. On the other hand, there are other instances in which the visible color is "worn" by an animal throughout its life, as happens with colors that serve to warn a predator of the dangerous nature of the animal they hope to catch. In order to be most effective, aposematic coloration must always be visible. A signal, on the other hand, must be undetectable by predators. The anolida lizard (above), found in Costa Rica, is capable of changing color, although in most cases its skin is a neutral brownish-grey. During the mating season, the males of some species inflate their throat sacs, which can be yellow or red in color, to attract the females.

In the case of this small
gecko of the Solomon islands,
the bright white markings
against the purple back-
ground could either be an
imitation of the dappled light
that penetrates forest canopy
or a means of warning
potential predators of the
lizard's dangerous nature.

Inside a complex labyrinth of coral, camouflage techniques differ from those found in the open sea or on land, because here, light breaks up into thousands of particles and the background is always extremely colorful. Fish whose appearance might seem to be highly distinctive in an aquarium become almost impossible to detect within their native environment. Among the techniques most often employed in protective coloration are those involving the breaking up of the animal's outline (left) and the subdivision of the body into numerous, indistinguishable small parts.

The large white spots on this crossbow fish (top), photographed in the waters of the Philippines, combine with the seemingly random other markings on the body to give the appearance of unconnected patches of pattern and color.

In South America, there are some very small frogs of the Dendrobates family called "poison-arrow" frogs. Secretions from the skin of these amphibians, when rubbed on the tips of the arrows of the Noamana, Choco, and Cuna tribes, can kill a large monkey. The colors of these frogs are among the brightest to be found among vertebrates. The almost phosphorescent yellows, blacks, reds, and greens contribute to make them truly striking creatures.

Poisonous snakes are often brightly colored; later on, we will discuss the American coral snake. The white, yellow, black, and red rings warn even the most distracted of predators. The only poisonous lizard, the Gila monster, and its relative, the Mexican heloderma, are black and red in color. Among birds and mammals aposematic colors are rare. This is perhaps because their predators are usually also mammals, whose methods of hunting are more often based on the sense of smell than on the sense of sight. Skunks, with their contrasting black-and-white stripes that warn of their "unprofitability," are an obvious exception.

A predator must learn to identify warning colors and then associate them with the potential dangerous nature of the prey. This is one of the reasons why these colors are so clear and easily recognizable. In contrast to mimetic camouflage, in which the predator does *not* associate the colors with the prey's presence, aposematic colors are used by the prey to *teach predators a lesson*. Conspicuous colors are easy to learn; the message is imprinted much more quickly in the minds of predators. Bright colors are also simpler to remember, or less difficult to confuse with others (these might seem to be similar hypotheses, but biologists make subtle distinctions between them).

Again, a predator's mental picture of its prey plays a part, but in an opposite way. Where lessons are concerned, those who teach need pupils who are are able to learn. And here is a small but curious detail: since it appears that insects do not have great learning capacity and only mammals and some cephalopods can remember the negative experiences to which they have been subjected, it follows that only animals whose predators are vertebrates and cephalopods need to develop warning coloration.

The colorful skins of coral fish are only useful during the day, when interaction with other members of the same or similar species creates the need to communicate. At night, it is best for the fish to disappear altogether. The parrot fish (above), for example, like many other members of the Scarida species, envelops itself in a mucous cocoon. Particles from the sea bed and other debris cling to the cocoon, helping to conceal the fish from its enemies. Pictured on the facing page is an Australian sea horse—a species which is extraordinarily well camouflaged. Numerous outgrowths resembling small algae have developed on its body. They even manage to fool another fish, which has sought refuge in the sea horse's "foliage."

Perhaps hundreds of millions of years ago, when insects were the only creatures to inhabit the land, the colors weren't as bright as they are today. Perhaps we owe the vast range of colors that surrounds us to the vertebrates who came before us.

If it is easy to understand why the bright colors of these species have evolved, it is not so easy to explain *how* they did so. Their usefulness is obvious, but if at the beginning (as is logical), there were only a few aposematic forms derived from mutations—a few more colorful butterflies, or one or two yellow-and-black wasps—the predators couldn't have known that catching and eating them could be dangerous. On the contrary, perhaps the colors attracted hunters even more, and the death rate of these specimens was higher than that of their well-camouflaged relatives. Perhaps the most acceptable hypothesis is that which suggests that these animals lived in groups and that the mutation which brought about visibility might have been shared among members of the group, with one or two sacrificing themselves for the common good.

Which animals benefit from being colored and visible? We spoke earlier about the common good; let us now look at the mechanism involved. If, in order to learn that a particular butterfly must not be caught, a bird must catch it at least once (and therefore probably kill it), the butterfly's colors have served no purpose if the insect becomes a victim. The ones who gain an advantage are the other butterflies of the same species, which, from that moment on, will be steadfastly avoided by the predator.

The genes responsible for the coloration present in the first butterfly, which were sacrificed in the bird's beak, are also present in its living relatives. A small part of the victim, therefore, remains in those who have profited from its death, because from that moment forward, the hunter will never touch them again. This is why we find aposematic animals in such large groups—whole societies of bees, wasps, and bumblebees, colonies of corals, groups of nudibranches, and immense swarms of Monarch butterflies, for example.

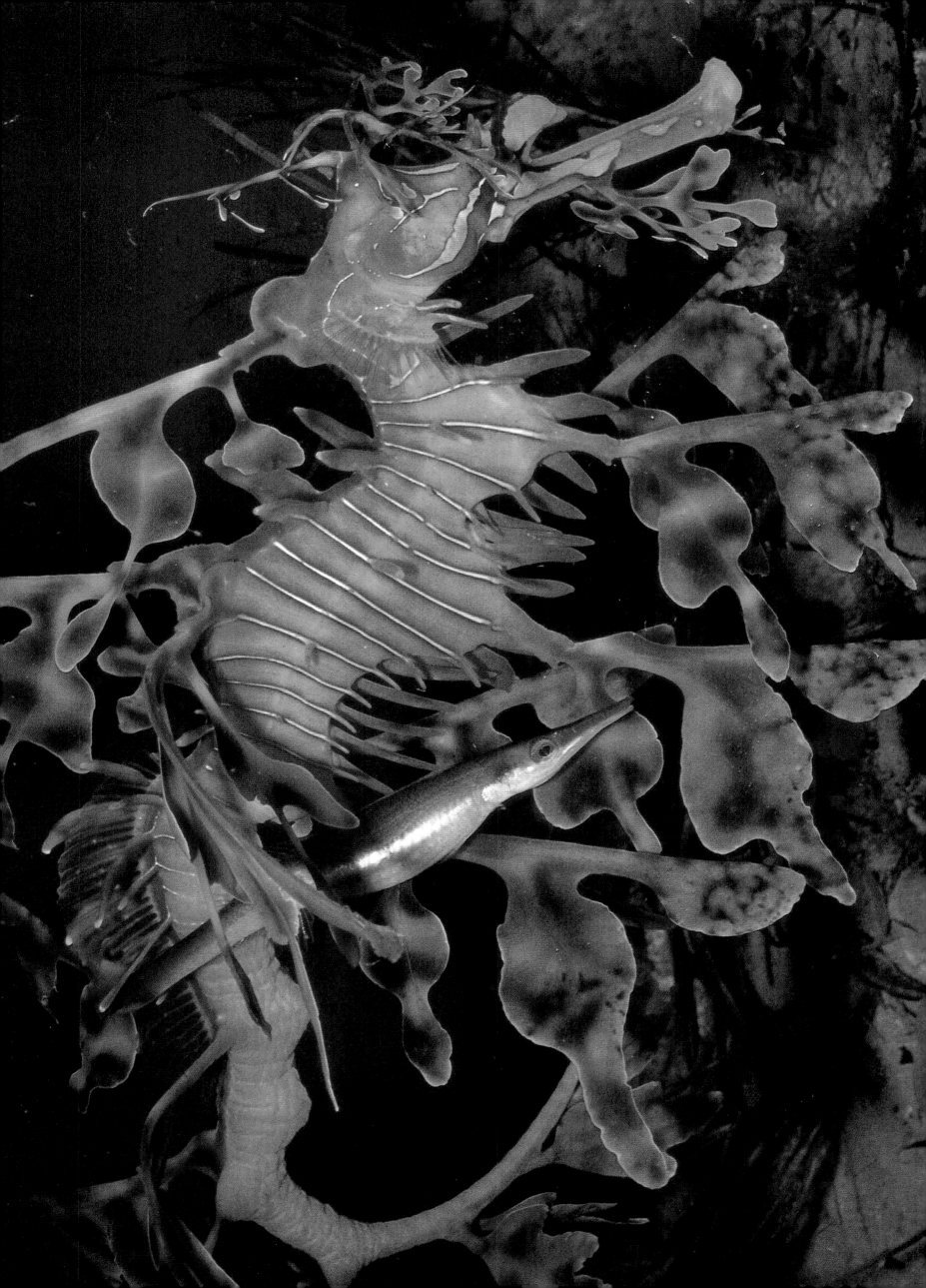

The nudibrancha are widely found in most of the seas throughout the world. Some of these mollusks, like the two species pictured here, have appendages on their backs that contain extensions of the digestive gland. Inside these extensions the animal accumulates "weapons," which it gets from other animals. In fact, some nudibrancha prey on coelenterates (sea anemones and coral polyps), whose tentacles contain numerous, small poisonous cells (stinging cells). The nudibrancha swallow these whole and later use them for protecttion against predators.

PAGES 118-119
A underwater cable covered with a startling array of species demonstrates the subtle distinction between plants and animals living in the seas. The red and green growths are sponges; the long, fernlike "trailers" are the spines of the hydroida.

This sea horse is an example of the many mechanisms of camouflage that can take place at the same time. The fingerlike protrusions help it to blend in with the algae that are so plentiful in the surrounding environment. The contrasting colors help to break up the outline of the animal's body so that a would-be predator cannot discern an overall shape. The "transparent" parts of the animal's face also serve as camouflage; they mimic the shadows and the spots of light that penetrate the expanses of algae among which the sea horse usually lives. All the members of the Singnatiform species, including the common Mediterranean sea horse, are characterized by slow movements; their almost complete immobility helps them to blend in with their surroundings.

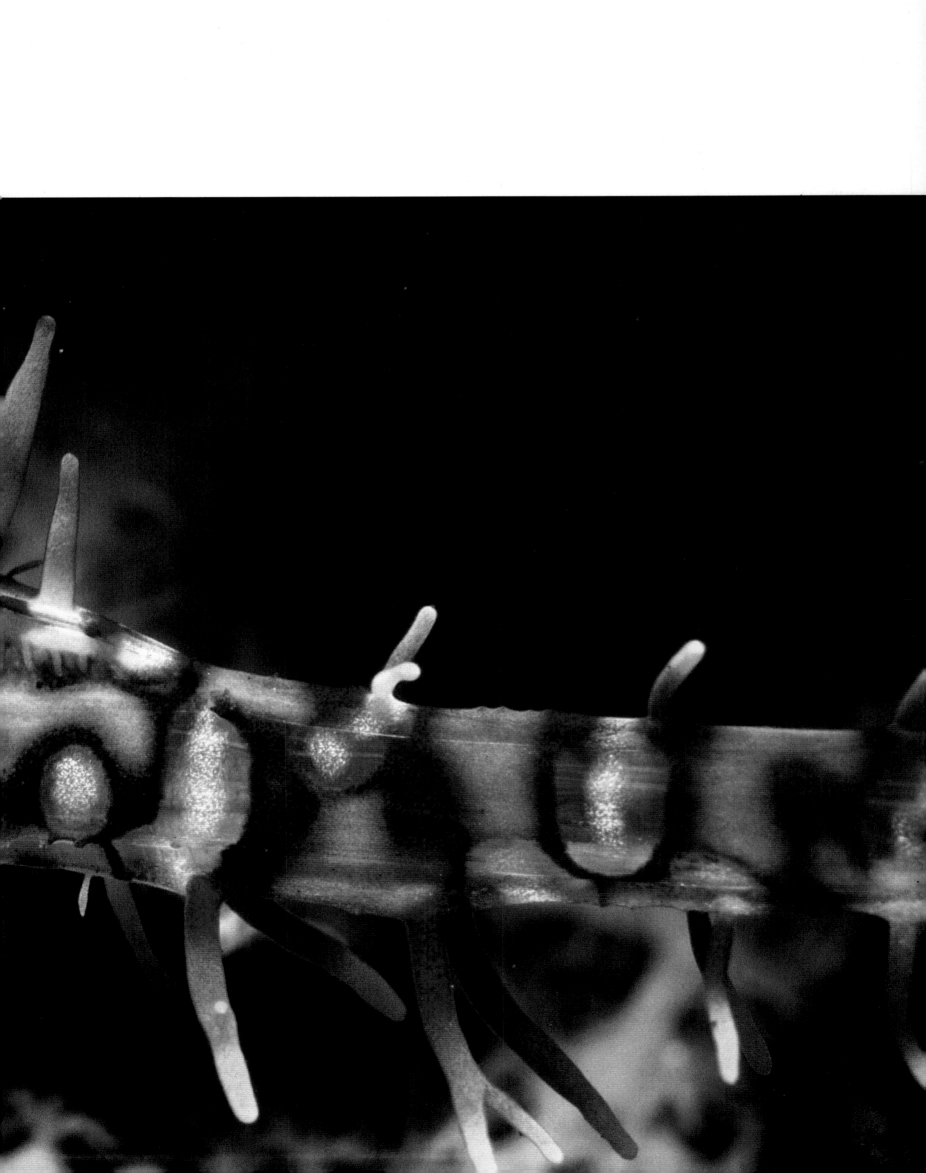

Scorpion fish (left) live along the sea bed in temperate or tropical zones and never venture out into the open sea. They are carnivorous and feed on smaller fish. The long, fan-shaped pectoral fins are an excellent deterrent to the fish's enemies, and the red-and-white stripes make it difficult for their prey to see them against a backdrop of coral. In the photograph below, only the eye of this small red fish resting on a coelenterate gives it away. Since the fins are the only parts of the fish that move, evolution has rendered them transparent so that they do not attract the eye of a predator on the prowl.

The Deceptive Imitators:
Batesian, Mertensian, and Muellerian Camouflage

Dr. Mueller, a zoologist working in the Amazon thirty years after Dr. Bates, asked himself, "Are groups of dangerous animals also made up of individuals from different species?" Greater protection could be attained when two or more species are poisonous, armed, or otherwise "not profitable" and live together in the same environment and share the same habitat. This is exactly what happens in what has come to be termed "Muellerian camouflage."

Dr. Mueller succeeded in explaining a phenomenon that Bates had come across but hadn't understood. We have already mentioned the similarity of wasps, bees, and bumblebees, all armed with harmful stingers. If several species are dangerous in the same way and adopt (by means of tiny mutations) the same skin, the danger of being caught is lessened by virtue of the sheer numbers of individuals. These "Muellerian chains" are very widespread, especially in the tropics.

The origins of Muellerian camouflage are easy enough to explain. Evolutionists talk of a "convergence" towards a color or a combination of colors, or a particular structure and stabilized selection. When many species adopt the same skin, Muellerian chains are created and the predators become wary. Moreover, Muellerian camouflage is quite a stable evolutionary system. An animal moving away from the basic models (even if only by a small degree) would find itself surrounded by predators ready to pounce on it. As in the case of cryptic camouflage coloration, the aposematic skins of Muellerian camouflage could appear to be perfectly integrated systems. The predators, who are aware of the danger, avoid the brightly colored prey and look for easier and more "profitable" prey.

When Henry Bates began to subdivide all the butterflies he had caught into taxonomically similar groups, he realized that he was making numerous mistakes. Those that, at a first glance, he had thought were Eliconidi, after careful analysis turned out to be members of the Pierids. Pierids are often preyed upon by birds. And here came another astonishing revelation: these butterflies are *imitators*—they take advantage of the fact that predators recognize and avoid certain color patterns.

For almost all the examples of warning coloration cited above, some species exist which are not poisonous, repugnant, or dangerous at all, but simply imitate animals that are. Bates stated in very clear terms that this type of variation "could only depend on natural selection, whose delegates are the insectivores, which slowly destroy those varieties that are not sufficiently alike (to their models) to trick them." Bates's hypothesis was immediately accepted by Darwin; it was exactly what he needed to confirm his own theory. He even wrote a favorable review of Bates's article.

There are insects that imitate bees; although they have two wings instead of four, Hymonoptera and some flies (Dipterans, especially Syrphides) are so similar to bees or wasps that they even trick people who catch them. Imagine how easy it is, therefore, to trick a bird that tries to eat them.

Ladybirds and other coleopters called Chrysomelids are protected by the liquids they secrete. In the tropics they are imitated by some species of cockroach, which copy both their physical appearance and their behavior. In the same way, some types of grasshoppers are identical (in the shape of their bodies and the positions they adopt) to another member of the coleopter family, which lives in the tropical forests of Southeast Asia. Once again, butterflies give us the majority of examples of this type of camouflage, which, in honor of its discoverer, is known as *Batesian camouflage*. Monarch butterflies are imitated by another species known as viceroys (*Limenitis archippus*) which, in most cases, are not poisonous but take advantage of the protection offered by the former.

Another even more extraordinary case takes place in the savannah of Brazil. Here, small black-and-yellow lizards (and even small snakes) imitate some species of equally striking millipedes whose poisonous secretions protect them from predators. It is one of few cases in which a vertebrate imitates an invertebrate. Only the young of this species (*Diploglossus lessonae*) imitate the millipedes of the Rhinocricus family. Animals that adopt the techniques of Batesian camouflage in color also often mimic the behavior of the animal being copied. The small Brazilian lizards that imitate the poisonous millipedes move (as opposed to their non-mimetic counterparts) with the same slow, deliberate movements of the millipedes themselves. The edible butterflies that imitate inedible ones fly in

Many of the caterpillars that become tropical butterflies have fascinating defensive and offensive weapons. One of the most extraordinary examples of mimetic ability is the development by these caterpillars of false eyes, which resemble the eyes of another, usually much larger and more dangerous species—in these instances, snakes. On one end of the bodies of each of these two caterpillars, one from Tanzania (above) and the other from Central America (right), are two large dark spots that look exactly like snakes' eyes. Whereas, in the caterpillar above, the eyes are simply black areas of color on either side, in the caterpillar on the right, the entire structure of a snake's head is almost perfectly imitated. There are even colors that evoke reflections of light in the "pupils," and the shape of the head is that of a poisonous snake. These caterpillars also imitate the snake's behavior: they shake the "head" and even hiss at potential predators.

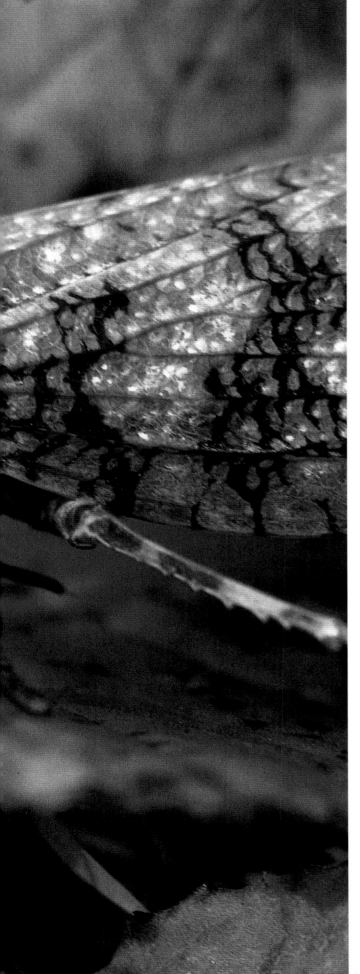

The fulgoridae are a species of large homoptera, often brightly colored, that live in the tropics. Their heads extend out in the form of a hollow "horn," which was once reputed to be luminous. The names of various species, including "lantern insects," is derived from this belief. The significance of these extensions is still not understood, although it is believed that their function may be related to courtship. Some extensions resemble small nuts, some resemble a crocodile's head, and others have no equivalents in nature. The hypothesis that the extensions act as deterrents has been refuted by some scientists, whose experiments have revealed that these insects were not at all frightening to insectivores. The camouflage of a fulgoridae homoptera in repose (top left) resembles the bark of a tree, although this particular insect has found itself in an area where its camouflage is of no use. When disturbed (top right), it opens its wings to reveal two large "eyes." Another species of fulgoridae (left) has a hornlike growth on its head that imitates the open mouth (complete with teeth) of a crocodile.

exactly the same way—with slow, deliberate movements. Predators can take as much time as they want to watch these beautiful flying creatures, and if they try to attack, the butterflies move even more slowly. This is quite disconcerting to the bird, and it is an effective means of defense.

Another example is found in one of few mammals which has, if not poisonous, at least repugnant flesh—the lemur. This small, primitive insectivore is always snubbed by predators. In Borneo at the beginning of the century, Shelford described a series of squirrels who lived with lemurs and took advantage of the protection the latter conferred upon them.

Cases of Batesian camouflage are less widespread in birds than in insects, but they do exist. Two small African flycatchers of the Stizorhina family are almost identical to other thrushes of the Neocossyphus family and are avoided by predators because of the pungent smell of formic acid, which they get from the ants they prey upon. When the wryneck (related to the woodpecker family) is disturbed, it mimics a snake's behavior, shaking its head from side to side and hissing. Some golden orioles from the Moluccas and New Guinea are only slightly similar in color to the five aggressive species of nectar birds with which they share their nesting areas.

Often it is simply a *model* of colors that is imitated and not necessarily in an exact order or arrangement—in other words, just enough to trick a predator.

Some species do not imitate the entire body of a predator, but only a part of it—the most dangerous or threatening part. Many caterpillars are perfectly mimetic, but when disturbed, they rise up on their legs and turn over on their backs, revealing two large "eyes" similar to those of a serpent. To enhance this imitation, they even emit a hissing sound. We could devote an entire chapter to the fascinating topic of false eyes. Some butterflies have "eyes" on the backs of their wings which they reveal only when threatened; a moment's indecision on the part of the predator is all the prey needs to escape. Animals are irresistibly attracted to and repelled by the eyes of other animals, and potential prey do all they can to hide their true eyes. They also do their best to show their false

These butterfly caterpillars (above) feed in groups. At the same time, they raise the posterior part of their bodies to reveal their bright yellow legs— perhaps in order to attract the attention of predators away from their heads— the most vulnerable part of the body.

eyes at the very last moment; false eyes are almost always large and frightening to predators.

The more accurate the eyesight is perceived to be, the better it is for the owner; some butterflies even have eyes on their wings which have a pupil, an iris, eyelids, and reflect light, bearing a striking resemblance to the eyes of owls. There is nothing more frightening to a small bird. But it is not only butterflies that have false eyes. Grasshoppers, mantises, and coleopters have them on their bodies: almost (as photographers Perennou and Nouridsany have described it) "an archetypal eye" that both frightens and attracts.

If the false eye doesn't necessarily fulfill its true function, a faithful Batesian imitator must follow some basic rules in order to achieve success. Fascinated by the hypotheses proposed on the subject of camouflage, many experts of Victorian England set about defending these theories from the attacks of critics. Let us not forget that it was at the end of the last century that the great debate between the supporters and detractors of the theory of evolution by natural selection took place. It was Bates's traveling companion, Alfred Wallace, who settled the matter. The hypothesis put forward by Bates must, according to Wallace, follow some basic rules in order to work.

First of all, it is obvious that the model and the imitator must live in the same area. Then, the imitator must be rarer than the model. If a predator encounters more harmless imitators than poisonous models, it will not be particularly shaken by the experience and will, especially if hungry, attack all animals who have that particular coloration. Up to this point, both the model and the imitator are equally at risk.

It is also in the imitator's best interest for the predator to have had more negative experiences than positive ones. If, however, the experience has been a particularly negative one, the effect of such experience will last for days. It is therefore better to imitate extremely poisonous species.

But from the model's point of view, things are not that simple. Being imitated by another species is not at all pleasant. If a spider imitates an ant, as happens in the tropical forests of Asia, the ant is at risk of being caught by a bird, who came across the spider—the imitator—first. Therefore,

Butterflies are among the most plentiful insects in the animal kingdom and they provide numerous examples of every type of camouflage. Some, especially moths, are particularly cryptic, i.e., they rely on camouflage to help them disappear into the background. Others—the poisonous or inedible species—are, in contrast, extremely visible, and warn their predators that catching them will be an unpleasant (and possibly deadly) mistake.

A papilionida extends its red, antennalike osmeterio (top right), an organ that some butterfly caterpillars release when they hear the enemy approaching. The movement of this brightly colored extension frightens small birds.

This large Australian butterfly caterpillar (bottom right) is covered with red and black stinging hairs.

even in this case, the same thing happens as in cryptic camouflage; the original model tries to "beat" the imitator by changing itself into something different and disengaging itself from the chain.

A poisonous or repugnant species cannot permit others who don't have these qualities to "ruin everything." The system of models and imitators is therefore derived from a continuous struggle to be protected from predators as much as possible, by means of either real or false weapons or any other methods that might work in any given situation.

If, in the area where a species is widespread, for example, there is no existing model to which the species can turn, that species tends to be cryptic. The butterfly *Bathus philenor*, which lives in North America, is aposematic and repugnant. In the area where it is most widespread, it is imitated by *Limenitis arthemis*. But in Canada, where the first species does not exist, the second has a completely different coloration. The *Papilio dardanus* has no species to copy in Madagascar, and both the males and females are cryptic.

Experts on camouflage believe that the situation is probably the following: one species "broke away" from the mimetic Muellerian chain. The coloration, and therefore the protection, remains the same, but the toxins are no longer collected and stored in the mimic's body. The trick, therefore, is at the expense of both predators and other species in the chain with similar coloration. However, it is obvious that some species must accept the responsibility of being poisonous; otherwise, all the logic behind camouflage disappears and the predators soon learn that the prey is harmless. Let us add one more small piece to the mosaic of camouflage.

In South America, in many diverse environments ranging from the humid jungles to the open savannahs and arid deserts, there are many poisonous and non-poisonous snakes that share the same vivid coloration. They are all covered with yellow, red, and black rings. They are all called coral snakes. Some of them, the true coral snakes, which belong to the Micrurus species of the Elapids family, are deadly. Others, which are called "false corals," are harmless or only slightly

poisonous. This would therefore seem to be a typical case of Batesian camouflage, in which the imitator benefits from the protection offered by the poisonous snake.

However, true coral snakes are so poisonous that an amazing phenomenon occurs. A predator that dared to catch a true coral snake would not simply be shaken by the experience, he would be dead. Therefore, the opportunity for learning a lesson is lost. Only the false, slightly poisonous or harmless coral snakes can teach the prospective predator the lesson. Therefore, the protection achieved is exactly the opposite of what the prey requires. The slightly poisonous, false coral snakes, are in fact the only models that harm the predator and send an "unprofitable" message. The true (and deadly) coral snakes and the non-poisonous ones benefit from the lesson.

In this case, the person who brought the matter to light was a famous German herpetologist by the name of Dr. Mertens, who studied coral snakes in South America. When more dangerous animals camouflage themselves to look like less lethal ones, the phenomenon is known as *Mertensian camouflage*.

We have come to the end of our journey into the world of colors. These tones, shapes and models all, in one way or another, possess a fleeting and yet tangible quality which is sometimes referred to as beauty or fascination. The profusion of colors of the butterflies, grasshoppers, fish, snakes, birds, and other creatures from around the world that have been captured in the photographs of this book captivate us and arouse our curiosity. Terms such as *cryptic, Batesian, Muellerian,* and *Mertensian* camouflage, interesting as they are to discover, do not adequately convey our deep fascination with the shapes and colors of the magnificent landscape of which we are all a part. These colors live within us.

Poisonous snakes are among the animals that have the best defense against their enemies—deadly venom. Many species that are not venomous take advantage of this fact by imitating them. One of the best imitative "circuits" is that involving coral snakes and their imitators (above). The red, yellow, and black rings are characteristic of the true (and deadly) coral snakes, the so-called "false" (nonpoisonous) coral snakes, and a second imitator that is mildly poisonous. The actual arrangement of the colored bands distinguishes one from the other. Interestingly enough, the mildly poisonous imitators do not benefit from the protection offered by the imitative coloration. This is because predators do not learn to avoid true coral snakes; the first and only encounter is lethal. However, both the true coral snakes and the false coral snakes benefit from the protection offered by the moderately poisonous imitators—the only snakes in the group who can teach the predator a lesson. An eyelash viper (right), which inhabits the tropical rainforests of Costa Rica.

Page 132-133
The whip snake of South East Asia lives in trees and is covered with green scales. The scales on the upper part of the snake's body are camouflaged to look like the patina of lichens and algae that cover the trees.

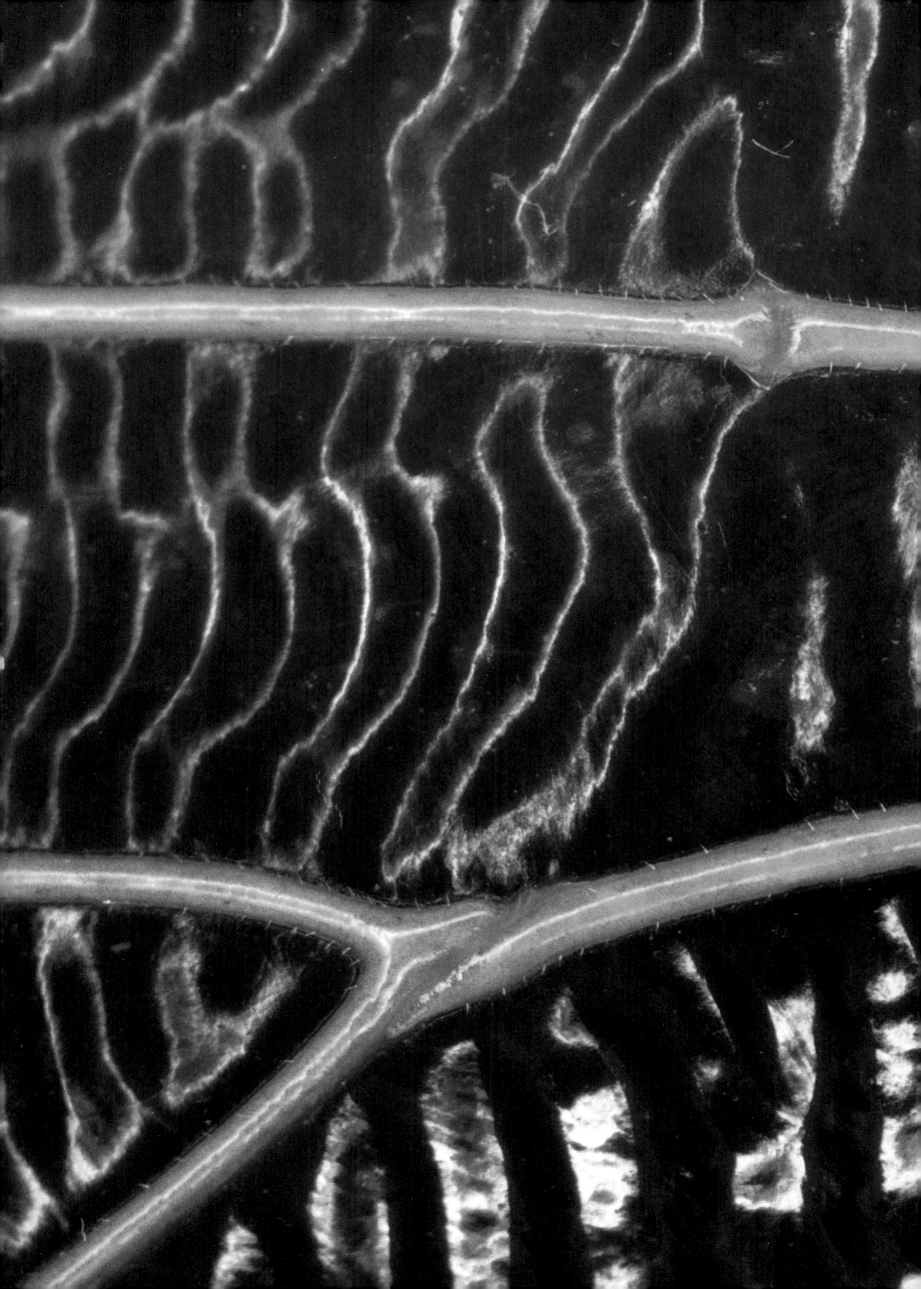

Photography Credits

Kurt Amsler: *pages 64 bottom, 65.*
P. Atkinson/Planet Earth: *pages 10-11.*
Liz Bomford/Ardea: *page 75 bottom.*
Anthony Bannister: *pages 40, 89 bottom.*
Jen and Des Bartlett/Bruce Coleman: *pages 70 top, 70 bottom.*
Marcello Bertinetti: *pages 60 bottom, 118-119.*
Bob and Clara Calhoun/Bruce Coleman: *page 91 bottom.*
Franz J. Camenzind/Planet Earth: *page 91 top.*
David Corke/Zefa: *page 124.*
G. D'Acunto/Panda Photo: *page 72.*
Sophie De Wilde/Jacana: *pages 12-13.*
Francoise Danrigal/Jacana: *page 116.*
Georgette Douwma/Planet Earth: *page 114.*
Alain Even/Diaf: *page 75 top.*
Jean-Paul Ferrero/Ardea: *page 107.*
K. Fiedler: *pages 42-43.*
Michael Fogden/Bruce Coleman: *pages 8, 30, 31, 76 top, 82-83, 84-85, 109, 110, 130, 131, 137 bottom, 140, 141.*
Paolo Fossati: *page 61 top.*
C. B. and D. W. Frith/Bruce Coleman: *pages 87, 132-133.*
Michael Gloer/Bruce Coleman: *page 61 bottom.*
Francois Gohler/Ardea: *pages 95, 127 top.*
Francois Gohler/Jacana: *page 126 top.*

Don Hadden/Ardea: *page 73.*
E. Hall/Ikan: *pages 62-63.*
Rudie H. Kniter/Ikan: *pages 56, 57, 58, 59, 104-105, 115, 123.*
Stephen J. Krasemann/Bruce Coleman: *page 125.*
Stephen Krasemann/Jacana: *pages 14-15.*
Waine Lankinen/Rapho-Grazia Neri: *page 1.*
Leonard Lee Rue/Bruce Coleman: *page 29.*
Kenneth Lucas/Planet Earth: *pages 60 top, 98.*
David P. Maitland/Planet Earth: *pages 129 top, 135.*
Jhon Mason/Ardea: *page 137 top.*
Fredy Mercay/Bruce Coleman: *page 129 bottom.*
Grazia Neri: *pages 16-17.*
Charlie Ott/Bruce Coleman: *page 90.*
Christian Petron/Diaf: *pages 120-121.*
D. M. Plage/Bruce Coleman: *page 106.*
Rod Plank/Zefa: *page 79.*
Fritz Polking: *pages 36-37.*
Eckart Pott/Bruce Coleman: *page 94.*
K. G. Preston/Mafham-Agenzia Masi: *pages 20-21, 22-23, 24, 26-27, 38, 44-45, 48, 49, 50-51, 52-53, 81, 89 top, 128, 134, 142-143.*
Andy Purcell/Bruce Coleman: *pages 32-33, 136.*
Luciano Ramires: *pages 92-93.*

Gary Retherford/Bruce Coleman: *pages 80, 108.*
Ed Robinson/Ikan: *back cover, pages 4-5, 99, 100-101.*
Carl Roessler/Planet Earth: *pages 64 top, 66-67, 112 top.*
Carl Roessler/Bruce Coleman: *page 112 bottom.*
Jeff Rotman: *pages 117, 102-103.*
Sauer/Bavaria: *page 126 bottom.*
K. Schafer/Zefa: *front cover.*
Peter Scoones/Planet Earth: *page 28.*
J. Scott/Planet Earth: *pages 34-35.*
Gerard Soury/Jacana: *page 54.*
E. Stella/Panda Photo: *page 86.*
Austin James Steven/Bruce Coleman: *page 9.*
K. H. Switak/Zefa: *page 78.*
Kim Taylor/Bruce Coleman: *page 41.*
Ron and Valerie Taylor/Ardea: *pages 96, 97, 122.*
Peter Ward/Bruce Coleman: *page 111.*
Art Wolfe/Zefa: *pages 2-3.*
Zefa: *page 69.*
Gunter Ziesler: *page 39.*
Gunter Ziesler/Bruce Coleman: *pages 138-139.*
Jim Zuckerman/Westlight/Grazie Neri: *pages 18-19.*
J. P. Zwaenepoel/Bruce Coleman: *page 68.*

Library of Congress Cataloging-in Publication Data

Ferrari, Marco, 1954-
   [Colori per vivere. English]
   Colors for survival : mimicry and camouflage in nature / Marco Ferrari.
     p.   cm.
   ISBN 1-56566-048-X
   1. Protective coloration (Biology) 2. Mimicry (Biology)
3. Camouflage (Biology) 4. Protective coloration (Biology)--Pictorial works. 5. Mimicry (Biology)--Pictorial works.
6. Camouflage (Biology)--Pictorial works. I. Title.
QL767.F42313   1993
591.57'2--dc20   93-22184
                                            CIP

Tropical forests are home to a large number of species of amphibians. The high level of humidity and the wealth of food and protective covering are factors that have led to a population explosion among these forest creatures. In ecosystems such as this, colors provide the perfect method of communication for small animals. This tree frog (below) lives within the abundant foliage of a rainforest in Central America. Its large red eyes serve as a warning to would-be predators. The small tree frog in the photograph at right, on the other hand, is more uniformly colored.

PAGES 142-143
The red markings on the body of this grasshopper mark the areas from which the insect can emit a harmful liquid when it is captured.

The haemolymph (body fluid) of insects is usually either green or colorless. In insects such as this, however, it is bright red or orange from the presence of carotenoids.

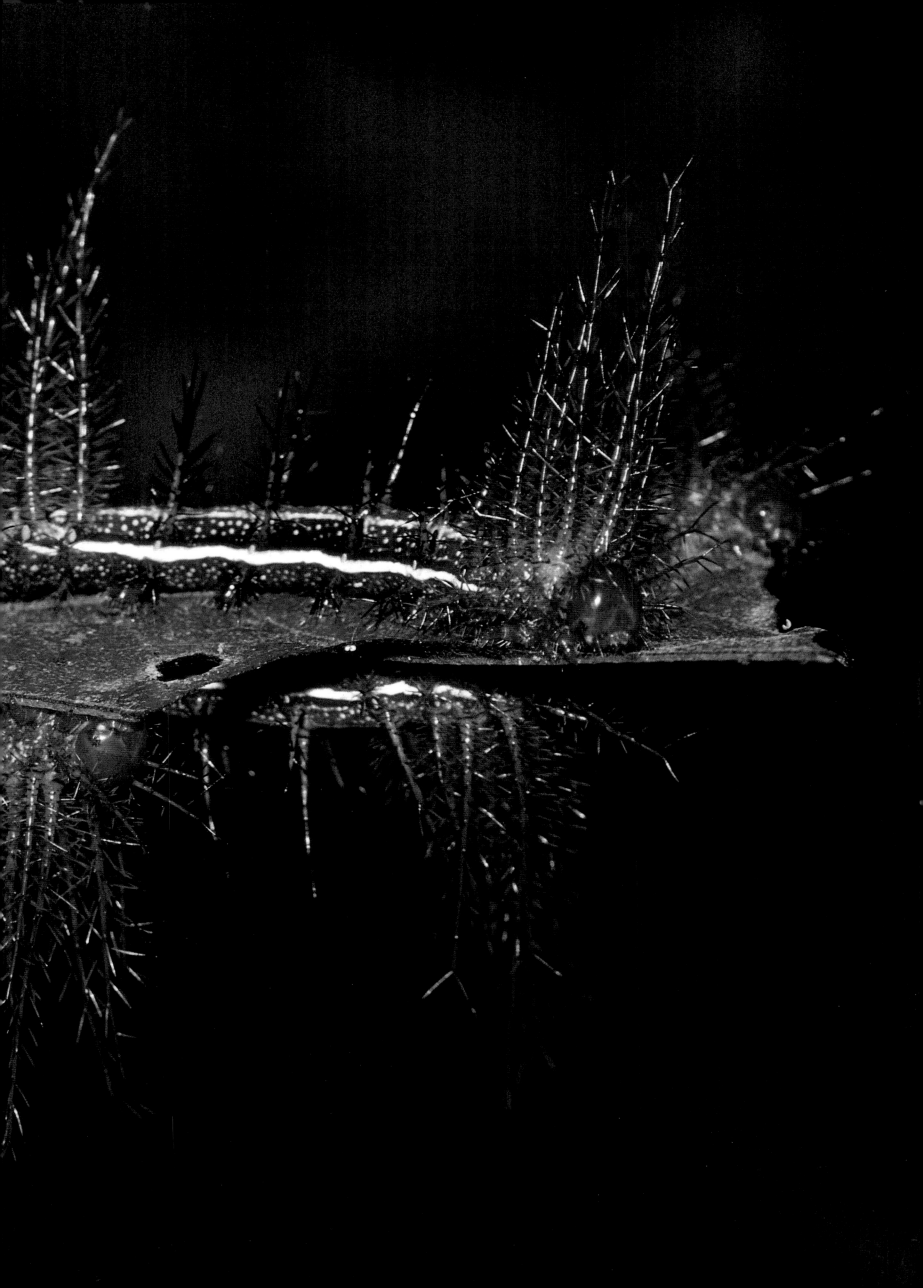

PAGES 138-139
A group of caterpillars rest on a leaf in a game preserve in Peru. This clustering behavior confuses predators; it is diffi-cult to distinguish a single caterpillar amid the tangle of poisonous thorny extensions.

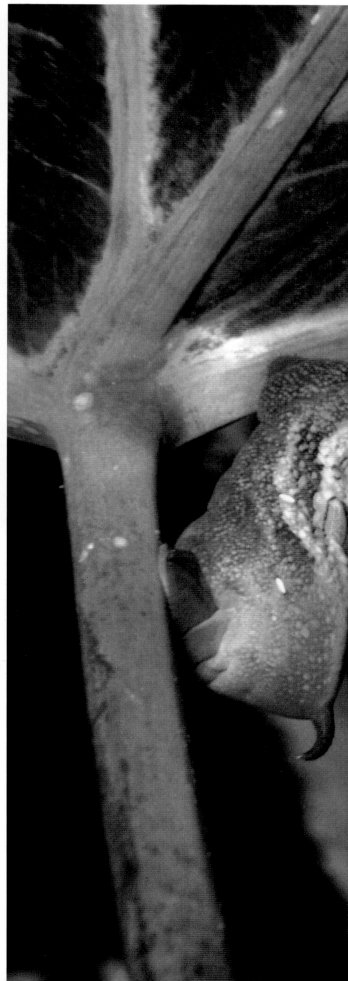

Butterfly caterpillars can defend themselves in many ways. The most commonly employed mechanism involves the use of special extensions along the sides of the body that will sting any predator trying to capture the insect. These extensions (above) are often brightly colored and contrast with the rest of the caterpillar's body. Another method of frightening predators is the thrashing around of a caterpillar's whiplike "tail." Many caterpillars are extremely sensitive to the presence of predators trying to catch them. Although a predator may pause for only a few seconds (unsure of his prey, given the defensive mechanisms), that moment's hesitation may be all that is needed for the caterpillar to escape, which it does more often than not by throwing itself to the ground. (Right) This Malayan butterfly caterpillar tries to trick and frighten away its enemies by displaying false eyes that imitate those of a snake.

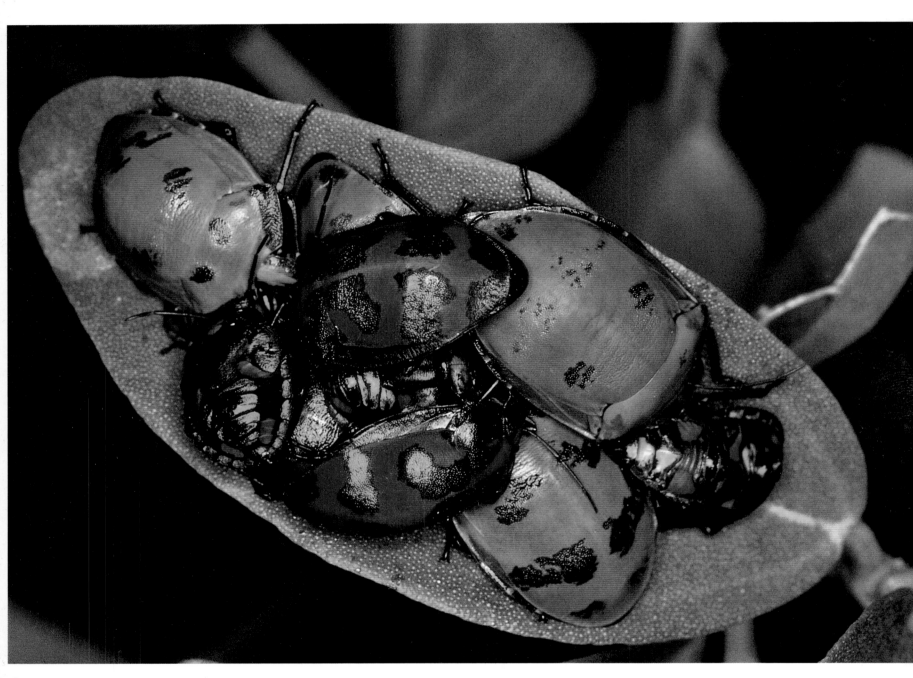

The harlequin bug uses two protective devices to ward off predators—bright color and an offensive odor.

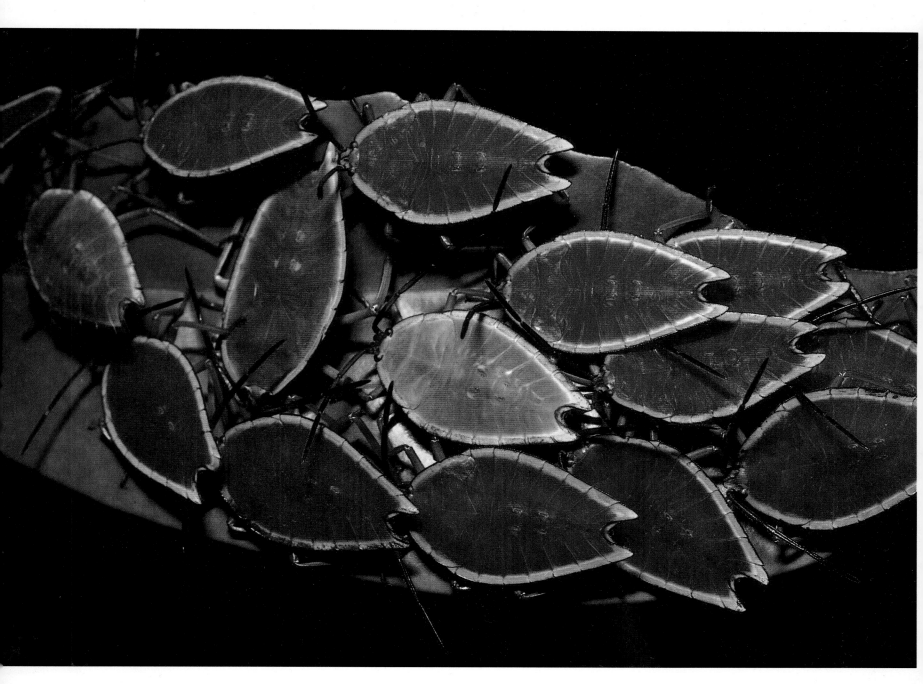

Insects provide some of the most clearcut examples of aposematic camouflage—that which emphasizes a creature's dangerous nature. These Australian insects live in groups during the nymph stage. The bright red color warns off predators. When an insectivore captures one of these bugs and has an unpleasant experience, others in the group benefit from the sacrifice of one of their own.